LIGHT

LIGHT

Michael I. Sobel

The University of Chicago Press
Chicago and London

Michael I. Sobel is professor of physics at Brooklyn College of
the City University of New York.

The University of Chicago Press, Chicago 60637
The University of Chicago Press, Ltd., London
© 1987 by Michael I. Sobel
All rights reserved. Published 1987
Printed in the United States of America

96 95 94 93 92 91 90 89 88   5432

Library of Congress Cataloging-in-Publication Data

Sobel, Michael I.
    Light.

    Includes index.
    1. Light.   I. Title.
QC355.2.S63   1987   535   86-25024
ISBN 0-226-76750-7

In memory of my father

Contents

Preface

A friend of mine once said that the trouble with being an architect is that everyone is an architect; everyone knows what's good and what's bad. I seem to remember the same story told by an economist. In physics we don't have that problem.

Generally speaking, people don't claim to know much physics, or much natural science altogether. True, things have changed a bit since C. P. Snow wrote of the "two cultures." Science curricula have appeared in the elementary school. Universities have gotten back to "distribution requirements," including some science. We now have science sections in the daily newspapers, a number of magazines aimed at a popular audience, and some fairly good programs on public television. (One attempt to put science on network television, even with the dean of television journalists as host, was a quick failure.) And yet, I am not sure how far we have come. A curriculum in the school doesn't ensure a teacher who uses it. A "core" course may be just an obstacle to the student, a bore to the instructor. As for the journalists, sometimes it seems that all they can provide is pop psychology ("Study shows marriages are better if husband has girlfriend"), dreamy speculations about the human mind, computers, chips, drugs, and bombs. And when the subject is deep and basic, we often read more about the professor's cluttered office and faded sweater than about his thoughts.

Yet to those of us who have had the pleasure to be immersed in it, science is more than a quick fix to the world's problems, and more than a collection of computer printouts, equations, and grant proposals. It is a grand and ennobling structure, like Notre Dame in Paris or like the Eroica, a tribute to the power and the breadth of human imagination. And just as the Gothic cathedrals soar toward the heavens and just as a theme runs through a symphony, so there is unity in the natural sciences. The great principles, like the second law of thermodynamics and the Pauli exclusion principle, and the broad classes of phenomena, like energy and waves, reach out and embrace the universe from the atom to the galaxies. An appreciation of these connections should be the privilege of every man and woman.

My objective in choosing the subject of *light* is to present it as one of these central themes and to follow it in many logical paths. These lead to the rest of the electromagnetic spectrum, to ultraviolet light, x-rays, and radio waves. They lead to the cosmic background radiation, to fiber-optic communication, and to the light-sensitive pigments in the eye. Unlike Notre Dame, science is not complete. I hope you will discover that and be on the lookout for the next step.

More years ago than I care to remember I received an award from the Woodrow Wilson National Fellowship Foundation for a project which included work on this book. This is the time to express my appreciation for that early vote of confidence. I am grateful also to Brian Schwartz, Joseph Krieger, and Alvin Halpern, who, in various administrative capacities, helped to provide time during which much of this book was written. George Skorinko, Brian Thompson, Dennis Hall, Charles Harris, Linda Mantel, Surendra Saxena, Melvin Belsky, Morton Glantz, and Joseph Krieger helped with conversation, advice, and insight. I am particularly indebted to Melvyn Selsky, Avigdor Ronn, Philip Morrison, Jane Gallagher, Harvey Lyman, and Israel Abramov, who read and commented upon portions of the manuscript. Naturally the final responsibility for what is written here rests only with me. Finally I would like to thank my wife, Carolyn, for encouragement and for caring, and, yes, for patience.

The Wave Theory of Light

1

Light fills our universe. In the deepest of night, in the farthest reaches of space, light is there. Yet we are strangely unaware of its presence. It may be that, for humans, light is so central to our perception that nature is forced to make us unaware of it. Processing light beams from all directions at all times would overload the circuits of the human brain. We interpret. We see not light but objects, constructed by the brain from information passed along the optic nerve. We construct shapes, colors, textures, and motion. Perhaps for this reason the ancient philosophers and thinkers did not devise elaborate theories of light equal in scope to the astronomy of Babylonia, the medicine of Egypt, or the algebra of India.

Often it is only the peculiarities of light's behavior—the distorted view of objects under water, the left-handed image in a mirror, or the play of sunlight on wavelets on a lake—that call our attention to the existence of something between an object and our consciousness of it.

Early Observations

Curved mirrors and convex lenses were used as far back as recorded history to concentrate the light of the sun and create fire. A lens of rock crystal was found in excavations at Nimrud, the royal city of Assyria. Indeed the earliest known writings about light have to do with *refraction,* the bending of a light ray in water or glass, and with *reflection,* the bouncing of a ray from smooth surfaces such as a mirror. The philosophers who followed Mo Ti in China in the fifth century B.C. studied optics and analyzed the images produced in plane and curved mirrors. Their approach to nature was empirical, accumulating laws governing the properties of light; they did not speculate in the manner of the Greeks. The earliest speculations about light are attributed to Pythagoras, who established a center of learning in southern Italy in the sixth century B.C. and gave the world some of its deepest insights in mathematics, philosophy, and mysticism. It may have been Pythagoras who proposed the idea that light consists of small particles sent out by objects toward the eye of the viewer; or this idea may have originated with the atomists (those propounding an atomic theory

1

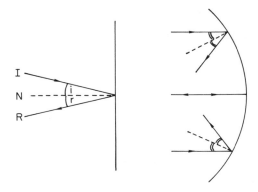

Fig. 1.1. The law of reflection. A ray of light, *I*, incident upon a mirror is reflected in the direction of the ray *R*. The dashed line *N*, is the *normal* (or perpendicular) to the surface. The angle of incidence, labeled *i*, is the angle between the incoming ray and the normal; the angle of reflection, labeled *r*, is the angle between the reflected ray and the normal. Angle *i* equals angle *r*. Even if the mirror is curved, there is at each point of the surface a normal direction (the dashed lines in the right-hand figure); the law *i = r* still holds at each point.

of matter) of his school a century later. An alternate hypothesis, that the eye sends out visual rays (as in the expression "Cast your eye upon it!") which then strike an object and cause the sensation of sight, has been attributed to Euclid, the great mathematician of the third century B.C. Euclid wrote two books on light, the *Optics* and the *Catoptrica* (meaning "on reflection"). The latter has been lost, but a review of it, written many centuries later—and possibly spurious—discusses the properties of mirrors and gives the correct law of reflection: a ray is reflected from a plane surface at the same angle with which it strikes the surface (fig. 1.1). Euclid was aware that light travels in straight lines, and on this assumption discussed the laws of perspective.

Euclid's century was the zenith of Greek science. Although Greek democracy was already in decline, the campaigns of Alexander the Great had spread Greek influence throughout Europe and Asia, and encouraged commerce, education, and science. The greatest scientist of the age was Archimedes of Syracuse, who made advances in mechanics, hydrostatics, and mathematics. Unfortunately his book on optics is lost, but it is known that he observed the change in appearance of an object as it sinks into water.

Claudius Ptolemy made the first systematic study of refraction. Ptolemy flourished in Greek Alexandria in the second century A.D. and his cosmology served the world for thirteen centuries. The Ptolemaic system, in which the sun and planets orbit the earth, was actually a compilation of work of earlier Greek astronomers, notably Hipparchus. But Ptolemy recognized that to determine correctly the location of a planet he must take account of the bending of light as it enters the earth's atmosphere. He showed that when light enters a dense medium from a less dense one, the ray is bent toward the *normal;* that is, toward the line perpendicular to the boundary between the media. (Conversely, it is bent away from the normal when entering a less dense medium). He demonstrated this with a simple experiment, undoubtedly known long before, in which a coin is placed at the bottom of a cup. The eye is so placed that the coin cannot be seen over the edge of the cup. But when the cup is filled with water, the coin becomes visible, because the light ray is bent at the interface of water and air (fig. 1.2). Similarly a light ray from the vacuum of interplanetary space is bent downward when it enters the atmosphere, and so a star or a planet appears to be higher in the sky than it is (fig. 1.3). Refraction of sunlight, in the same way, raises the apparent position of the sun and increases the duration of daylight by five to ten minutes.

Ptolemy studied air-water, air-glass, and water-glass interfaces, and made tables of the *angle of incidence* (the angle between the incoming light ray and the normal) and the corresponding *angle of refraction* (the angle between the refracted ray and the normal). He stated (incorrectly) that for a given interface the two angles were proportional. His *Optics,* only part of which has survived to the present, is remarkable not so much for its data as for the very fact of its adherence to the experimental method.

Yet the next major figure in the story of light is found more than eight centuries later, during the flowering of learning under Islam. By the year A.D. 750 the soldiers of Mohammed had built an empire along the entire Mediterranean Sea and as far east as India. In the ensuing times of prosperity the caliphs tolerated the traditions of classical learning in centers in North Africa and the Middle East, thereby encouraging science and philosophy. Among the great thinkers of this period, al-Khwarizmi the mathematician, Avicenna the physician, Averroës the philosopher, and Alhazen the physicist, it is the last who commands our attention.

Abu Ali al-Hasan ibn al-Hasan ibn al-Haitham, born in 965, became known in the West by the Latinized form of his name, Alhazen. In his writings on optics we find not only highly developed applications of algebra and geometry but an experimental approach in a very modern sense, with the use of sighting tubes, strings, and plane and curved glasses and mirrors to study the

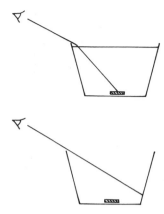

Fig. 1.2. A penny at the bottom of a cup of water is visible because light from it is refracted at the water-air interface. With the eye in the same position and water removed, the penny cannot be seen.

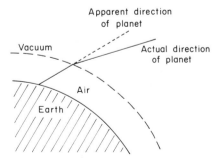

Fig. 1.3. Light rays entering the earth's atmosphere from the vacuum of outer space are bent downward. Thus the light seems to come from a point in the sky higher than the actual source of light. (The path of the ray does not really make a sharp angle at the top of the atmosphere. Since air gradually thins as we reach the top of the atmosphere, the light path makes a smooth curve.)

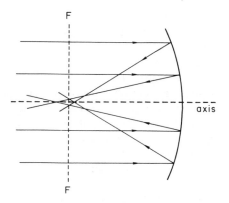

Fig. 1.4. Spherical aberration. Parallel rays reflected from a spherical mirror fail to converge at a single point. Incident rays farther from the axis converge nearer to the mirror. In plane *F* the image is fuzzy. (In this and other diagrams of optical instruments, the angles that light rays make with the horizontal axis—and with each other—may be exaggerated for the sake of clarity.)

laws of reflection and refraction. Alhazen showed that Ptolemy's claim that the angle of incidence was proportional to the angle of refraction was incorrect, although Alhazen himself did not obtain the correct mathematical relation. He observed that a convex lens, each surface being a part of a sphere, had the power of magnification. He carried out a complete mathematical analysis of reflection from spherical, cylindrical, and conical mirrors. He showed that if parallel rays of light strike a curved mirror that is a section of a sphere, the rays are not brought to a precise focus, a condition now called *spherical aberration* (fig. 1.4); but a mirror in the shape of a paraboloid (a solid figure produced by rotating a parabola about its axis) produces a sharp focus.

Alhazen considered twilight, the persistence of daylight after the sun has set, to be due to the refraction of sunlight from the upper layers of the earth's atmosphere. Making the assumption that twilight ends when the sun's rays are refracted from the very top of the atmosphere, he deduced that the atmosphere was twenty to thirty miles high, a good estimate by modern standards.

In Alhazen's writings we find the most detailed early dissection of the human eye. We find speculation on the method of propagation of light which presages the seventeenth-century theory of Huygens. And we find the suggestion, demonstrated only in the nineteenth century, that light flows less easily (i.e., at lower velocity) when it enters a dense medium, and that this causes the ray to be bent toward the normal.

The research and scholarship of the Moslem scholars were important not only in themselves, but also because they transmitted, in Latin translation, the spirit of learning to those medieval philosophers like Roger Bacon and Albertus Magnus whose work foreshadowed the age of science in the West. Bacon was well acquainted with Alhazen's optics, understood how to trace rays of light through lenses and mirrors, and may have been the first to use a lens for spectacles. He also suggested the combination of two lenses to make a telescope, although it is not certain that he actually constructed one.

We pass swiftly by the age of Copernicus, the invention of the microscope by Zacharias Janssen about 1590, the invention of the telescope by Hans Lippershey in 1608, and its quick deployment by Galileo Galilei. Some time before 1621, the Dutch scientist Willibrord Snell discovered the correct law of refraction. Snell's law gives the relation between the angle of incidence and the angle of refraction in terms of a quantity, called the *index of refraction,* which is characteristic of the medium into which light travels (fig. 1.5). Modern wave theory shows that it is the ratio of the speed of light in air to the speed of light in the medium.

Not long afterward, the French mathematician Pierre de Fermat unified the laws of reflection and refraction by showing that both could be deduced from the postulate that light chooses a path of least time. That is, given two points *A* and *B* in a region with mirrors or with different media, the path of a ray from *A* to *B* will be that for which the time of travel is least. Implicit in Fermat's principle is the assumption that light travels at a finite speed (so that travel time is a meaningful concept) and that the speed is lower in a denser medium. The fact that light travels at finite speed was not demonstrated until after Fermat's analysis, but the question was considered open at the time (and discussed by Galileo).

The idea that a physical system behaves according to a minimization or a maximization principle finds its way into theoretical physics as an economical and profound formulation of scientific law. A ball rolling on an undulating surface comes to rest at the lowest point; a chain hung between two points shapes itself so as to minimize a quantity called potential energy; several fluids at different temperatures, when mixed, will transfer heat among themselves in such a way that the entropy, a measure of disorder in the system, is maximum. The least-time principle for light had been hinted at earlier by Alhazen and others. It impressed Fermat as a deep manifestation of the simplicity of nature.

Newton and Color

Into the history of light now enters Isaac Newton (fig. 1.6), perhaps the most influential scientist of all time. Newton's greatest contributions were his general laws of motion and his theory of universal gravitation. Building upon the earlier researches of Galileo, Newton's laws allowed one, in principle, to calculate the future motion of any material body, given a knowledge of the forces acting upon it and its condition at some initial instant. He proposed, moreover, that there exists a universal force of attraction, gravitation, between any two bodies. Using this force to-

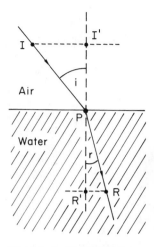

Fig. 1.5. Snell's law of refraction. A light ray travels from air into water, making an incident angle, i, with the normal to the water surface. The ray travels into the water making a refracted angle, r, with the normal. Snell's law is expressed in terms of the sine of angle i and the sine of angle r. The ratio $(\sin i)/(\sin r)$ is equal to a quantity n, called the index of refraction. The index of refraction is a constant; it is a number independent of the incident angle. Snell's law determines the angle r. The index of refraction is characteristic of the medium that light enters. For water, $n = 1.33$; for glass, $n = 1.5$ (although there are different values for different types of glass); for diamond, $n = 2.42$. For a vacuum, $n = 1$, and for air, n is slightly greater than 1. (The discussion above neglects the difference between air and a vacuum.) In the figure $\sin i$ is equal to the ratio $\overline{II'}/\overline{IP}$, where $\overline{II'}$ stands for the distance from point I to point I' and $\overline{IP}$ stands for the distance from I to P. Similarly, $\sin r$ equals $\overline{RR'}/\overline{RP}$.

gether with his laws of motion, he was able to give a quantitative explanation of such diverse phenomena as the falling of an apple from a tree, the flight of a stone across a field, the orbit of the moon around the earth, and the orbits of the planets around the sun. Newton's laws, first published in 1687 in his *Principia Mathematica,* unified heaven and earth and demonstrated—contrary to the views of the ancients—that the same laws of physics apply on earth and in outer space.

Less far-reaching, but nonetheless marking a turning point in the history of science, were Newton's researches into the nature of light, carried out around 1666 when he was a young man at Cambridge University. It had been known since antiquity that when sunlight is bent in a glass prism the colors of the rainbow are produced, but there was no satisfactory explanation for this observation. It was suggested, for example, that the spectrum was produced because the sun is more than just a point of light; that light rays from different parts of the sun strike the prism at different angles and as a result emerge from the prism at different angles and with correspondingly different colors. Newton's first step was to demonstrate that the angular size of the spectrum produced by the prism was too great to be accounted for by the finite size of the sun.

Imagine a darkened room, a small hole in the window shade admitting a narrow pencil of sunlight (fig. 1.7). The beams of sunlight diverge as they cross the room and produce a small circle of light on the opposite wall. The young scholar places a prism in front of the hole and observes the beams, which now strike a different area of the wall. The displacement of the beams is not surprising; it is the result of refraction of the entire beam by the

Fig. 1.7. Sunlight enters a room through a small pinhole. It is refracted by a prism and produces an elongated spot on the wall opposite. The angles are defined by drawing lines from the pinhole to the edges of the spot. Along the short (horizontal) axis of the spot the lines make an angle of 0.5 degrees. Along the long (vertical) axis, the angle is 3 degrees.

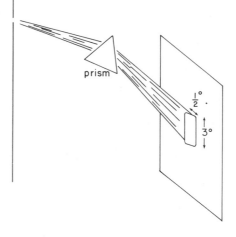

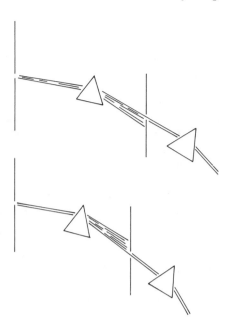

Fig. 1.8. Dispersion of white light. After a prism breaks white light into the colors of the spectrum, a barrier with a small hole is introduced, to select in turn red, violet, or inter-mediate colors. A second prism shows that violet light is refracted more than red.

prism. But the spot of light is now elongated; it is no longer a circle. Colors are spread out along its long axis, red on one end, violet on the other.[1]

Newton carefully measured the angle subtended by the axes of the spot as seen from the hole and found the long axis close to 3 degrees and the short axis about half a degree. It is easy to verify that the spreading of half a degree can be accounted for by the different angles of incidence of the sunlight—because, again, of the finite size of the sun. But how can the image spread to 3 degrees along the long axis of the spot? Newton proceeded to ana-lyze the beam, keeping the prism in place and introducing beyond it a barrier with a small hole. Through this hole he allowed to pass only the red portion of the beam; later the orange, the yellow, and so on. He then could experiment separately with each color, and he did so by placing a second prism beyond the barrier to refract light arriving through the small hole (fig. 1.8). Newton observed the angle of refraction of red light alone, orange light alone, etc., and found that the angle is different for different colors; violet light is bent most, red least, the other colors falling in between. Thus his theorems 1 and 2: "Lights which differ in colour, differ also in degrees of refrangibility [refraction]," and "The light of the Sun consists of rays differently refrangible."[2]

In another experiment Newton synthesized white light. He took the beam emerging from a prism—separated into colors— and let it pass through a second prism, inverted with respect to the first. Thus the second prism performed the opposite function of

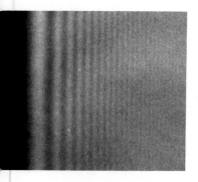

Fig. 1.9. The *diffraction* pattern produced at the edge of an opaque barrier. The barrier is on the left side of the figure and its edge is vertical. Light striking the barrier does not produce a sharp shadow, with dark on the left and light on the right. Rather the shadow edge is fuzzy; some light "bends" into the region where shadow is expected. With an intense light source we can see a pattern of alternating light and dark stripes at the edge. (Courtesy Brian J. Thompson)

the first: it combined the separated colors and produced a white spot on the wall.

Convinced that sunlight is a combination of the colors of the rainbow, Newton was then able to explain the colors of material objects. Long before Newton, scientists had argued about whether color is an integral part of a colored object or is associated with reflection of light from the surface of the object. Using a prism, Newton took light of a particular color ("uncompounded light," as he called it) and let it shine on different colored objects. A red object in blue light looks blue; in green light it looks green; and in red light it looks red, but in this case it appears brightest. Thus he concluded that a red object reflects all colors but reflects red more strongly, so that in the "compounded" white light of the sun the object reflects red most strongly and appears red to the viewer.

Newton's *Opticks* presents a wide variety of experimental researches, with illustrations, mathematical analysis, and deduction of general laws. We find very little theory, little speculation concerning the underlying nature of the phenomena. Newton thought of light as similar in some ways to a stream of particles. Light travels in a straight line, like a bullet. Light casts sharp shadows, so that, if we illuminate a barrier in front of a white screen, the region of the screen behind the barrier is completely dark and the region outside the barrier completely lit. Light, in other words, does not go around corners.

Or does it? In 1665 there were published posthumously the studies of the Jesuit Francesco Grimaldi on the nature of shadows. Grimaldi carefully observed the shadow of a very small circular obstacle placed in the path of light and showed, first, that the shadow was slightly smaller than it would have been if light traveled precisely in a straight line past the edge and second, that around the shadow were fringes of color. In other experiments he observed a slight fuzziness and separation of color at what should have been the sharp boundary between light and shade (fig. 1.9).[3] Newton knew of this work and carried out similar experiments himself.[4] He also studied the colors formed in light reflected from thin films such as insect wings. He observed the phenomenon, now known as *Newton's rings*, in which the convex surface of a lens is placed on a plane glass surface and a series of concentric light and dark circles is seen in the reflected light (fig. 1.10).

All these effects are now recognized to be consequences of the wavelike character of light. But although there are wavelike aspects in some of Newton's conceptions of light, his later writings emphasize the particle (or corpuscular) theory. His eminence was so great that the corpuscular theory held sway into and throughout the eighteenth century.

The Wave Theory

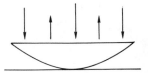

To understand the importance of the wave-particle controversy as it later developed, we must step back and consider the present position of wavelike phenomena in the structure of physics. Consider, for example, the V-shaped wave that spreads out behind a motorboat moving through the waters of a lake. You sit in your canoe watching the wave approach you, speed constant, wave height constant, knowing that when it hits you the canoe will be rocked back and forth and perhaps will capsize. But what is it that "hits"? What is it, in fact, that moves across the water at that constant speed? Energy surely is transported across the surface of the lake, since it takes energy to rock your canoe, and that energy is produced by the motorboat. But water is not transported across the surface: the water that rocks the canoe is the same water that lay quiet under you a moment before. The motions of the water are only up and down. First the surface of the lake is raised in a small region near the motorboat; a little later that surface is flat and the neighboring surface is raised; and in this way an influence, a disturbance of the water surface, moves across the lake. (Such waves are called "surface waves.")

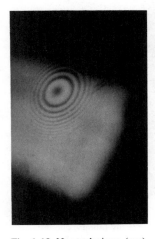

Fig. 1.10. Newton's rings. (*top*) Light shines on a lens resting on a flat glass surface and is reflected upward. (*bottom*) Viewed from above, a pattern of alternating light and dark concentric circles is seen. (Courtesy George Skorinko)

Thus scientists distinguish two methods by which energy can be transported from one place to another: by the transport of matter (I may ram your boat with the bow of mine) or by means of a wave. Another kind of water wave is the series of circular ripples produced when a stone is dropped into water. The ripples are formed by many alternating crests and troughs moving radially from the stone's point of impact. It is similar to the spherical sound wave emanating in all directions from a loudspeaker.

When energy is transported it is not always easy to tell whether the mechanism of transport is matter or wave. But there is one fundamental difference, a difference that becomes the crucial factor in the history of the study of light. Referring again to the wake of a motorboat, suppose two motorboats pass in opposite directions, setting up waves that cross one another. When the crest of one wave moves across the crest of the other, there will be a double influence for raising the height of the water surface, hence, momentarily an extra high crest. But when the crest of one crosses the trough of the other, the two influences—one to raise the water surface and the other to lower the water surface—cancel each other (partially), and the surface is only weakly disturbed. This is called *interference: destructive interference* if two waves cancel (or subtract) and *constructive interference* if two waves add. Interference is the phenomenon that distinguishes waves from matter. If two boats are located in the same space, they do not cancel each other; they are both really there. But a wave is an in-

Fig. 1.11. Christian Huygens (Burndy Library; courtesy AIP, Niels Bohr Library)

fluence—more precisely, an influence which travels—and two opposite influences may destructively interfere with each other.

The earliest systematic wave theory was put forward by Newton's great contemporary, Christian Huygens (fig. 1.11), of The Hague. Although not fully developed mathematically, Huygens's theory showed how a train of waves striking the edge of a barrier will be bent around the edge; thus a wave of light might penetrate into the region behind a barrier (fig. 1.12). Huygens was also able to explain Snell's law of refraction, if he assumed that waves travel more slowly in a dense medium than in air (fig. 1.13).

The bending of light around an edge is known as *diffraction*. And although it had been observed by Grimaldi and Newton, the amount of light bent into the shadow is so small that Newton did not regard the effect as evidence for light waves but rather as some property of the edge. In particular, the propagation of light in this respect may be compared with that of sound, which quite obviously goes around corners. Someone calls you from the next room; you hear her voice around the doorpost, but you do not see her. When the optical researches of Newton (1672) and Huygens (1678) were published, it was well known that sound was a pressure wave in air. It was not until the research of Thomas Young (fig. 1.14) in 1801 that the wave theory of light was revived and the similarities as well as the differences between light and sound were explained.

Young was considered by some to be the last man who knew everything. He was a physician, and, while waiting for his prac-

Fig. 1.12. Light passing the edge of an opaque barrier. (*a*) Huygens's view of light as a wave. Each point on a wave front acts as a source for a new spherical wave. For an uninterrupted wave (*left*), each succeeding wave front is a plane parallel to the earlier front. If a wave comes upon the edge of a barrier (*right*), the succeeding wave fronts bend into the region behind the barrier. The shadow behind the barrier does not have a sharp edge. (*b*) Light as a stream of particles. Those particles that hit the barrier are blocked; those that miss the barrier travel straight ahead. A sharp shadow is produced on a screen beyond the barrier.

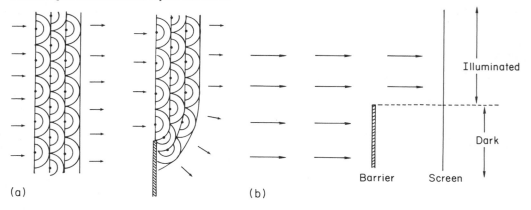

tice in London to develop, he found time to pursue researches in physics and mathematics and to occupy the chair of Natural Philosophy at the Royal Institution. At various points in his career he also contributed to the fields of botany, philosophy, languages, and physiology, and he made an important contribution to archeology by translating ancient Egyptian hieroglyphics using the Rosetta Stone as a key. In 1801 he presented to the Royal Society of London a lecture entitled "The Theory of Light and Colors." He undertook to prove the wave theory of light by demonstrating interference effects and to give the first physical analysis of the nature of color. For example, in the experiment of Newton's rings (fig. 1.15), Young argued, there exists a thin layer of air between the convex lens and the flat surface beneath it. Light is reflected from any boundary between two media, and in this case he considered light reflected respectively from the upper and the lower boundaries of the air layer. Two such beams arrive at the eye having traveled paths of different lengths. Thus they may arrive with a crest of one moving together with a trough of the other, destructive interference—a dark ring. A little farther from the center of the lens the path difference (between the two reflected rays) is slightly longer, and we find a crest of one wave moving with a crest of the other, constructive interference—a bright ring.

A given wave train of light in Young's analysis is a series of crests and troughs, and the difference between constructive and

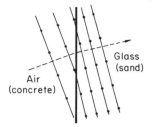

Fig. 1.13. Huygens's explanation of refraction. Huygens assumed that light travels more slowly in a dense medium (water or glass) than in air. Imagine a series of parallel wave fronts impinging obliquely on an air-glass interface. Or think of an analogy: ranks of marching soldiers crossing a boundary from concrete to sand. Each dot in the figure represents a soldier, or a point on the wave front. Each series of dots connected by a line represents a rank, or a wave front, and these fronts are shown at successive times. Soldiers on the right flank enter the sand first and are slowed down; soldiers on the left are still marching on concrete, moving faster; and so the whole rank veers to the right. By the time the rank has entered the sand completely, it is traveling in a direction closer to the normal. The light "ray" is simply the line drawn along the direction of travel. Thus it is bent toward the normal when light enters a dense medium—a medium in which it travels more slowly. Conversely, a ray is bent away from the normal when light leaves a dense medium and enters air.

Fig. 1.14. Thomas Young (Courtesy AIP, Niels Bohr Library)

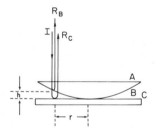

Fig. 1.15. Interference in Newton's rings. An incident ray, I, is reflected from the lower surface, B, of the lens. It is also reflected from the flat surface, C, of the glass plate. The two reflected rays, R_B and R_C, interfere. The nature of this interference, whether it is destructive or constructive, depends on the height, h. (The difference of path length for the two rays is twice h.) In turn, h varies with r, the distance from the center.

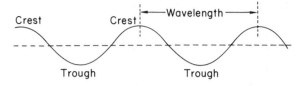

Fig. 1.16. The definition of wavelength

destructive interference depends on the distance from crest to trough; or, if we define the *wavelength* as the distance from one crest to the next (fig. 1.16), the difference between dark and light depends on a path difference of one-half the wavelength. Thus, by observing the separation between the light and dark rings, Young was able to calculate the wavelength of light of different colors. He found a value of about 0.00007 centimeters for red light, about 0.00004 centimeters for violet light, and values between for the intermediate colors in the spectrum. At a stroke Young penetrated many of the secrets of light. He explained Newton's rings and a number of similar effects associated with thin films, such as the colors seen in soap bubbles or in thin layers of oil floating on water. He himself performed the simplest interference experiment, allowing a light beam to fall on two parallel slits, and observing the pattern of light and dark stripes, often called *fringes,* on a screen beyond (figs. 1.17–1.19).

Young's findings establish the true nature of color, showing that each color in the prism's spectrum corresponds to a light wave of a particular wavelength, with the longer wavelengths toward the red end of the spectrum and the shorter wavelengths toward the blue and violet. Sound waves and light waves are expected to be very different, because the wavelengths of light are so extremely small (on a human scale) and the wavelengths of sound are large. (Typical values for sound range from 15 centimeters for a high-pitched whistle to 3 meters for a deep male voice.) Here lies the explanation of why the wave character of light is elusive: because the wavelength is so small, wave effects tend to be very small and hard to detect, unless the dimensions of the apparatus used are very small. Interference fringes are seen only when the spacing between two reflection surfaces is very small. Diffraction is seen easily only in the shadow of a very small object. Light travels around corners, but the path of a light beam deviates from a straight line by a very small angle.

One other property, the *polarization* of light, discovered in 1669, played an important part in shaping the wave theory as it developed through the nineteenth century. Certain transparent crystals have the property of affecting light passing through them in a peculiar way. Take, for example, a thin slice of the mineral

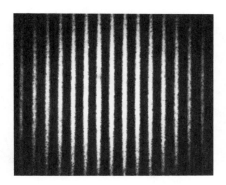

Fig. 1.17. Light intensity recorded on a screen behind two small holes that are illuminated by a single small light source. The dark stripes are places where the light waves from the two sources interfere destructively; light stripes are places where light waves interfere constructively. (Courtesy Brian J. Thompson)

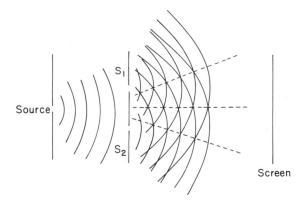

Fig. 1.18. Young's explanation of the two-slit experiment. Light from a small source passes through the two slits, S_1 and S_2. The curved lines represent crests of the emerging waves. At certain points crests of the two sources coincide and there will be constructive interference. Dashed lines are drawn through these points. In the field beyond the sources, there is constructive interference along certain *directions*. In those directions bright bands will be produced on the screen. Between those directions are the regions of destructive interference, and dark bands.

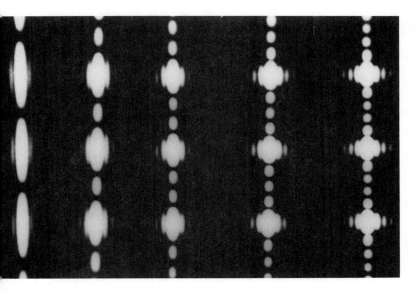

Fig. 1.19. Generalized examples of the two-slit experiment. With three, four, five, and six slits, more complicated interference patterns are recorded. (Courtesy Brian J. Thompson)

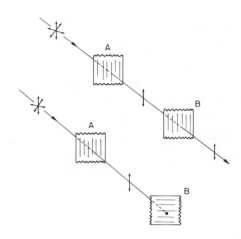

Fig. 1.20. Polarization. A polarizing crystal, *A*, allows through only that part of an incident beam that is polarized in a particular direction, say, parallel to the crystal's axis. Polarized light enters a second crystal, *B*, which passes the light if the crystal's axis is aligned with the axis of *A* (*upper figure*). If *B* is rotated 90 degrees, it does not pass the polarized beam (*lower figure*).

Fig. 1.21. Oscillations in a light wave traveling perpendicular to the plane of the page—out toward your eye. In the figure on the top, light is unpolarized: oscillations occur in all directions in the plane of the page. In the figure on the bottom, light is polarized: oscillations occur in only one direction.

tourmaline, often used as a gem, and pass a ray of light through it (fig. 1.20). Now let the transmitted ray pass through a second slice of the crystal. Keep the first crystal still while rotating the second crystal about an axis along the path of the light. As you rotate this crystal you find that at a certain position there is *no* transmitted light. Rotating the crystal from this position, you find that the intensity of transmitted light increases and it reaches a maximum after a rotation of 90 degrees. What is suggested by this and other experiments done earlier is that the propagation of light is, in some basic sense, a three-dimensional process. Suppose the beam of light is traveling horizontally through a windowpane toward the eye. The two slices of crystal are parallel to the plane of the window, one nearer than the other. We can understand polarization by assuming that the oscillations making up the light wave are in the directions parallel to the windowpane (perpendicular to the beam) (fig. 1.21). In ordinary light the oscillations can be in any direction in this plane and presumably vary randomly in all possible directions around a circle. But the tourmaline crystal is asymmetric and allows only those oscillations in, say, the up-down direction to pass through. The light emerging from it is polarized, meaning that the oscillations are entirely (or preferentially) along this up-down axis. Human vision is not sensitive to polarization, so this polarized light looks the same to us as ordinary light. But now pass the polarized light through a second crystal. If the axis of asymmetry of this crystal is parallel to that of the first crystal, then it allows through all of the up-down oscillations of light; the strength of the beam is not diminished. But if the second crystal is rotated 90 degrees, then the incoming polarized light is along the "wrong" axis and it is not transmitted; the strength of the beam is diminished almost to zero.

Thus, although polarization is a subtle property of light, it was instrumental in establishing one of the basic characteristics of light waves and an important difference between light and sound: light is what is called a *transverse wave,* the oscillations being perpendicular to the path of propagation. Sound waves in air, on the other hand, were known in the nineteenth century to be *longitudinal,* consisting of oscillations of air in directions back and forth, along the direction of propagation.

Although Thomas Young's crucial analysis of light did not immediately carry the day, it gradually came to prevail during the 1820s and 1830s, largely through the efforts of two French scientists, Augustin-Jean Fresnel and Armand-Hippolyte-Louis Fizeau (see fig. I.1). Fresnel made extensive studies of diffraction, measuring the fringes produced at the edges of shadows in various types of experiments (figs. 1.22 and 1.23), and he showed that his results could be fully explained by the wave theory, with wavelengths of light in agreement with those calculated by Young.

Fizeau measured the speed of light and showed that it was slower in water than in air. This was considered at the time to be a crucial experiment. Corpuscular theory assumed that light travels faster in a dense medium, the argument being that the molecules of the medium attract the particles of light, speeding up their motion as they enter; and then when light leaves the medium the same attractive force pulls the corpuscles back, slowing them down as they enter the air. The wave theory, according to arguments going back to Huygens, requires a slower speed in a dense medium than in air.

The wave theory was a product of efforts on the two sides of the English Channel. The respectability of that theory was attested to by the election of Fresnel, in 1827, as a foreign member of the British Royal Society and the election of Young, in the same year, as a foreign member of the French Academy.

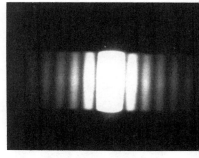

Fig. 1.22. Diffraction pattern of a single slit. In an experiment of the type elaborated by Fresnel, light strikes a single slit in an opaque barrier. Instead of seeing a single strip of light, we see a diffraction pattern. There is a "central maximum," the bright region in the middle, flanked by dark strips and secondary bright strips. This pattern can be thought of as a consequence of interference between light waves arising from different parts of the slit. (Courtesy Brian J. Thompson)

Electromagnetic Waves

Yet we may feel at this point a certain quality of never-never land about the description of light waves. For although we have made

Fig. 1.23. Diffraction pattern of an opaque strip. Instead of seeing a single dark strip, the shadow of the barrier, there is a dark central region, actually slightly smaller than the barrier, surrounded by characteristic diffraction fringes. (Courtesy Brian J. Thompson)

Fig. 1.24. Michael Faraday
(Courtesy AIP, Niels Bohr Library,
E. Scott Barr Collection)

analogies with waves in water and with sound waves, nowhere in the discussion of light is there any assertion of what it is that waves, what it is that oscillates in such a way as to establish the traveling influence. Light waves travel not only through water, glass, and air but through outer space, where there is, presumably, no matter at all—through a vacuum. Indeed, to the scientists of the nineteenth century the notion of waves in a vacuum was inconceivable. Schooled in the properties of water waves and sound (which travel through air and through liquids and solids), scientists felt it necessary to postulate the existence of a medium that carried waves of light as transverse oscillations. This was the *luminiferous ether,* which pervaded all space out to the stars. The properties of the ether were marvelous and paradoxical. It was rarer than air, since starlight is refracted when entering the atmosphere; it was incompressible, or else it would dissipate a light ray in all directions; it was perfectly elastic, so that it would not retard the passage of matter (such as the planets in their orbits). Even more strangely, it possessed the kind of rigidity that a solid has, because only a solid is able to propagate a transverse wave.[5] Lord Kelvin (see fig. 5.2) likened it to a mold of transparent jelly.

Philosophy in the nineteenth century was ripe for a concept like the ether, because scientists had come to accept the existence of substances other than ordinary matter. Science in the previous century, under the shadow of Isaac Newton, had been dominated by a mechanistic world view in which all phenomena were due to the interaction of bits of matter. But research into the properties of heat, light, electricity, and magnetism made it clear that although one might try to picture these as materials—fluids or corpuscles—they were nonetheless in some fundamental ways different from ordinary matter. Thus there were thought to be two types of matter: normal, or "ponderable" matter, and the "imponderables." The ether took its place then as the most pervasive and subtle of the imponderables. The great achievement of the nineteenth century was to unify the imponderables and to do so by founding the theory of light upon the phenomena of electricity and magnetism.

This great theoretical synthesis was the work of James Clerk Maxwell, but it grew out of the experimental researches of many scientists, including Hans Christian Oersted in Denmark, Jean-Baptiste Biot, André-Marie Ampère, and others in France, and Michael Faraday (fig. 1.24) in England. That there was an intimate connection between electricity and magnetism was first shown in 1820, when Oersted observed that a coil of wire carrying an electric current produces the same kind of magnetic field as a bar magnet. Faraday demonstrated the converse phenomenon, electrical effects produced by magnets (fig. 1.25). Although holding a magnet in the vicinity of an electric wire does not induce any

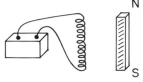 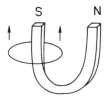

Fig. 1.25. Electricity and magnetism. In the figure on the left a battery produces electric current in a coil of wire. The coil then acts as a bar magnet. For example, it will attract small bits of iron. In the figure on the right a coil of wire is moved rapidly upward, through the magnetic field of a permanent magnet. An electric current is produced in the wire, although the wire is not attached to a battery or any other source of electricity.

flow of electric current, if a magnet is *moved,* for example, through a coil of wire, then a pulse of electric current is produced (without the wire being attached to a battery or any source of electricity).

Maxwell (fig. 1.26) was an important figure in many areas of nineteenth-century science (including thermodynamics, the statistical theory of gases, and the theory of color vision), but his greatest work was to bring together the various studies of electricity and magnetism and to create the unified theory which came to be called *electromagnetism.* His theory concentrates on the links between the two phenomena and in particular on the dynamic nature of those links; that is, they require motion (or some change over time) for their manifestation. Thus a changing magnetic field creates in its vicinity a changing electric field; and a changing electric field similarly creates a changing magnetic field. These two fields can thus be regarded as manifestations of the same underlying phenomenon—electromagnetism. Maxwell also deduced that they can exhibit wavelike properties (fig. 1.27).

For example, an oscillating electric charge will create around it a region of oscillating electric and magnetic fields; and these fields in turn create oscillating electric and magnetic fields around themselves; and so there proceed outward from the original charge radiating spheres of electromagnetic fields. Because of the periodic nature of the motion of the electric charge, the variation of the field out along a radius is also periodic. It is a spherical wave, analogous to a wave that travels along a plucked string or to the circular waves that radiate from the point where a stone is dropped into water. (The waves on the string are distributed over a one-dimensional space, the waves on the water surface over a two-dimensional space, and the electromagnetic waves over a three-dimensional space.) In these cases the quantity that is "waving" is the height of the string or of the water above a natural resting level. In the case of electromagnetism the quantity that is waving is the electromagnetic field. It is not a thing, not a piece of matter, but an influence—or a potential—something that *would* act if there were an appropriate receiver for its action. A tree falling in the forest will produce a sound wave, but it will not be heard unless an ear is present to process the wave. An electric charge or a small magnet placed in the path of an electromagnetic

Fig. 1.26. James Clerk Maxwell (Courtesy AIP, Niels Bohr Library)

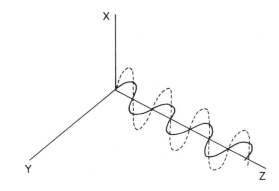

Fig. 1.27. A view of a polarized electromagnetic wave at a fixed instant of time. The light ray travels along the axis marked *Z*. The dashed curve represents the magnetic field; it points either up or down, in the *X-Z* plane, and its value oscillates as we move along *Z*. The solid curve represents the electric field; it points either left or right, in the *Y-Z* plane, and its value oscillates similarly. At a later time the whole pattern will have moved along the *Z*-axis, its shape unchanged.

wave is an appropriate receiver; it will respond by "feeling" a force and by going into a forced oscillation itself. (See app. A.)

A scientific theory is only as good as its quantitative predictions. The theory of electromagnetism, published first in the *Philosophical Magazine* in 1862, predicted one important number, the velocity at which electromagnetic radiation travels. In Maxwell's words, "This velocity is so nearly that of light, that it seems we have strong reason to believe that light itself (including radiant heat, and other radiations if any) is an electromagnetic disturbance in the form of waves propagated through the electromagnetic field according to electromagnetic laws."[6] Coincidentally, in the same year as Maxwell's publication, Jean-Bernard-Léon Foucault, improving on the earlier work of Fizeau, determined a new, accurate value of the speed of light (differing by less than 1 percent from today's best value).

Electromagnetic theory predicts that an oscillating electric charge will generate an electromagnetic wave and that the wave travels at a unique speed. The wavelength depends on how rapidly the charge oscillates. If the *frequency* of oscillation (number of oscillations per second) is high, then the wavelength is small, since the emerging wave pattern repeats itself very quickly. Picture a ribbon of cloth emerging from a sewing machine. If the sewing needle moves up and down very quickly (a high frequency of oscillation), then the stitches are close together (a short wavelength).

We often use frequency as a means of characterizing electromagnetic or other kinds of waves. Think of the wave as it passes by an observation post. The frequency of the oscillator that originally set up the wave is also the rate at which wave crests pass by the post—that is, the number of crests passing per second. Thus frequency is a property not only of the source but also of the wave itself.

From this picture it is not difficult to see that for light (or any wave) a simple relation exists among the three quantities frequency, wavelength, and speed. If ten complete waves pass by an observer in 1 second, the frequency of the wave is 10 cycles per second. If the wavelength of that wave is, say, 2 meters, then it is clear that a wave train 20 meters long has passed the observer in that 1 second. Thus the wave's speed is 20 meters per second. Speed equals the product of frequency times wavelength. And as stated above, high frequency means short wavelength; low frequency, long wavelength. Red light with a wavelength of 0.00007 centimeters, or 700 *nanometers,*[7] corresponds to a frequency of 4.3×10^{14} oscillations per second; violet light, at 400 nanometers, corresponds to 7.5×10^{14} oscillations per second.

The Electromagnetic Spectrum

Even before the time of Maxwell the spectrum had been extended beyond the violet-to-red range revealed by the prism. In 1801 the astronomer William Herschel, discoverer of the planet Uranus, made careful studies of the heating effects of light. He found that as he moved a thermometer across the beam emerging from a prism, from yellow to orange to red, heating effects increased and that they increased still further when he moved to a region beyond the red. Thus there is radiation beyond the red, called *infrared,* which, following the work of Maxwell, we recognize as electromagnetic waves with wavelengths longer than 700 nanometers. Herschel showed that infrared radiation obeys the same laws of reflection and refraction that light obeys. Later workers found effects of polarization and interference in infrared.

At the opposite end of the spectrum, *ultraviolet* radiation (waves beyond the violet) was discovered by the extension of another property of light. It had been known since early in the eighteenth century that light affects silver nitrate and certain other salts, causing them to darken. This phenomenon later became the basis of photography. In 1801 the German physicist Johann Ritter, having learned of Herschel's infrared waves, looked for and found a darkening of silver nitrate by waves in the region beyond the violet. He thus discovered what we would call electromagnetic radiation with wavelengths shorter than those of visible light.

In other words, we begin to understand light as a special case of a much broader classification of phenomena, the electromagnetic wave, which encompasses radiation of all different wavelengths. A narrow band, from 0.00004 to 0.00007 centimeters, is detectable by the human eye, and so has been known since antiquity as "light." We may call this the *visible* part of the spectrum. But other parts of the spectrum (according to observations made of sunlight) manifest themselves in more indirect ways. Ultraviolet light is invisible, but it affects the skin and causes sunburn. Infrared is invisible but produces the sensation of heat. Hold your hand a foot or so above the burner of an electric

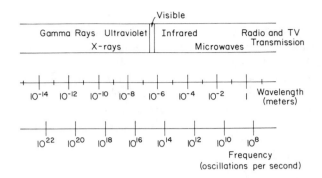

Fig. 1.28. The electromagnetic spectrum. These different types of radiation are all examples of electromagnetic waves with different wavelengths and frequencies.

stove. The burner glows red, as the hot coil emits visible light (not the same spectrum as the sun's, however), and you feel the heat, as the coil emits infrared. Now turn off the burner. As it cools down the red glow soon vanishes, but your hand still feels the heat: infrared radiation from the still-hot coil.

We thus can see Maxwell's work in a very broad context, providing an explanation not only for electricity, magnetism, and light, but more generally for a vast range of electromagnetic wave phenomena beyond the visible, encompassing ultraviolet and infrared and, as later discoveries would show, radio, x-rays, and gamma rays (fig. 1.28). Yet we should beware of looking at history with the arrogance of hindsight. Not all of the implications of Maxwell's theory were apparent in the period immediately following his work, and the depth of his accomplishment was not immediately understood.

For the sake of clarity we have discussed the generation of electromagnetic waves by an oscillating electric charge, but this treatment of the *source* of waves, although implicit in Maxwell's theory, was not addressed in his publication. Rather, he was concerned with showing that the interaction of electric and magnetic fields was such as to allow the *propagation* of waves. We may observe the tension in a violin string and thus see that it can carry a wave; a mechanism, such as a bow, for generating the wave is another matter. Maxwell's paper itself was somewhat inaccessible because of the difficulty of his mathematics, and rival theories of electricity and magnetism remained current, particularly in Europe. Nineteenth-century physics, influenced still by a mechanistic view of matter, felt the need of a laboratory demonstration of electromagnetic theory.

The Experiment of Heinrich Hertz

In 1871 a major center for research in electricity and magnetism was established in Berlin under the direction of Hermann von Helmholtz, then the leading figure in German science. Helmholtz encouraged his most gifted student, Heinrich Hertz (fig. 1.29), to investigate electromagnetic phenomena, and Hertz took up the study of rapidly oscillating electric charges. He constructed a device consisting of a brass rod about 1 foot long with a small air gap at the midpoint, across which he produced an alternating spark. Because of the sudden and violent surge of electricity in a spark, the motion of charges corresponds to oscillations at very high frequency (several hundred million oscillations per second). The sparks in the gap induce an alternating current in the rod, and the rod becomes the equivalent of a transmitting radio antenna. According to Maxwell's equations this antenna should send out waves of electromagnetic radiation with a wavelength twice the length of the rod.

Hertz's crucial experiments were carried out in 1888, some years after his time in Helmholtz's laboratory, when he held a teaching position at the Technische Hochschule in Karlsruhe. Imagine the demonstration hall, completely darkened to allow adaptation of the eye, the spark gap antenna at one side of the room. At the other side is a receiver designed by Hertz, a similar brass rod with an adjustable gap in the middle. When the transmitter is turned on, an invisible electromagnetic wave crosses the room, inducing an alternating current in the receiver. A faint glow is observed in the gap of the receiver, indicating sparks produced by the induced current. This transmission of man-made waves across many meters of empty space is dramatic proof of electromagnetic radiation as predicted by Maxwell.

By measuring both the wavelength and the frequency of these electric waves, Hertz was able to calculate the speed at

Fig. 1.29. Heinrich Hertz
(Courtesy AIP, Niels Bohr
Library)

which the waves travel, finding it to be precisely the speed of light. He also demonstrated many properties analogous to those of light, including reflection, refraction, and interference, and thus was drawn to the inescapable conclusion that light and electromagnetic waves are one and the same phenomenon.

Hertz did not believe that his "electric" waves would find any practical application. But in the early 1890s a young Italian inventor, Guglielmo Marconi, read of Hertz's work and imagined communication by Hertzian waves. Marconi's first wireless message was sent in 1895. Hertz died of cancer at the age of thirty-seven in 1894, but in his short career he achieved one of the major milestones in human history.

Having reached this culmination of nineteenth-century science, we must recognize that a vast gap exists between Hertz's waves and visible light. The frequencies of the former are about 5×10^8 oscillations per second;[8] frequencies of light are about a million times greater. The great quest arising from the success of Maxwell and Hertz was to find a mechanism in nature that can oscillate at more than 10^{14} times per second, rapidly enough to produce light. It was natural to look for this mechanism in the properties of the individual (and still hypothetical) atom. More than thirty years would elapse before the atomic theorists of the twentieth century would unravel the mystery thus posed.

Optical Instruments

2

Take an ordinary magnifying glass, the kind used for reading small print, and hold it near a wall across the room from a brightly lit lamp. If you slowly vary the distance between the lens and the wall, there will be some point—perhaps 8 or 12 inches from the wall—at which you will see a clear image of the lampshade. Look closely and you will see that it is inverted. Just such an image lies upon your retina when you look directly at the lamp. It too is inverted, but your brain interprets it as right side up.

Picture in your mind the rays of light proceeding toward you from the lampshade. From each point on it, light radiates in all directions, with rays from all parts crisscrossing each other as they move across the room toward the magnifying lens. Out of this entanglement the simple curved glass constructs its image, employing only Snell's law of refraction.

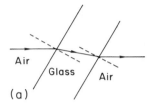

(a)

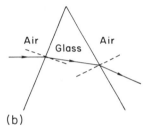

(b)

Formation of an Image

Suppose a ray of light enters glass at an angle; that is, it is not perpendicular to the glass surface (fig. 2.1). In the glass it is bent toward the normal (the perpendicular to the surface). Let the ray pass through the glass and reach the opposite surface, where it will again be refracted, this time bending away from the normal. If the second surface is parallel to the first (as in an ordinary windowpane), then the second refraction just undoes the effect of the first, and the ray emerges in the same direction that it had before it entered the glass. But if the second surface is not parallel to the first, as in a prism or near the edge of a convex lens,[1] then the emerging ray deviates from the direction it had before it entered the glass. If a series of rays all travel parallel to the optical axis of a lens (the line perpendicular to the lens through its center), those rays entering near the top are bent downward; those entering near the bottom are bent upward; and rays passing through the center, where the two lens surfaces are parallel, are unchanged. The result is that these rays are all brought together on the far side, and meet at a point, called the *focal point* (fig. 2.2).

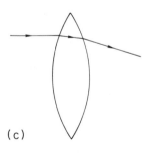

(c)

Fig. 2.1. Refraction of light: (*top*) through a plate of glass, (*middle*) through a prism, and (*bottom*) through a convex lens.

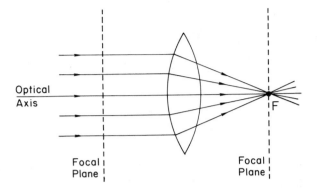

Fig. 2.2. A convex lens. Light rays parallel to the optical axis of the lens converge at point *F*, the focal point of the lens. Shown in the diagram is a cross section through the lens. In three dimensions the lens is formed by rotating the figure about the axis. Each surface of the lens is then a segment or cap of a sphere.

Now consider the rays emerging from one particular point on the lampshade and approaching the lens (fig. 2.3). These rays are not parallel; they diverge slightly as they enter the lens. They may still be brought together on the far side, but they will be less convergent and will meet at a point beyond the focal point. If rays are traced in this way from several points on an object, the rays from each point meet at a corresponding point on the far side of the lens. This *set* of points of convergence is called a *real image*

Fig. 2.3. Formation of a real image by a lens. (*a*) Rays of light from a very distant object are almost parallel and so converge close to the focal point. (*b*) Rays of light from a nearer object converge at a point farther behind the lens. (*c*) For an extended object, we follow three rays from, say, the head of the arrow on the left and see that they meet at a particular point on the right—the head of the inverted arrow. Similarly, three rays from the tail of the real arrow meet at a point on the right—the tail of the inverted arrow. The inverted arrow is a real image.

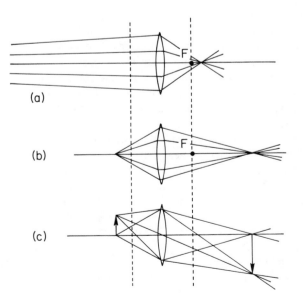

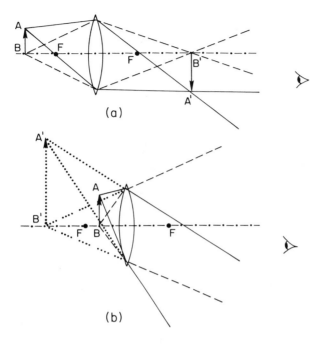

(a)

(b)

Fig. 2.4. A real and a virtual image. (*a*) The real image is formed as in fig. 2.3*c*. There is no material body at the image position, *B'* to *A'*; but a viewer on the far right would receive the light rays diverging from that image and interpret them as if there were an inverted arrow at the position of the image. (Remember that the angles these light rays make with the axis may be greatly exaggerated in the figure. To view this image, we accept only those rays that enter the pupil of the eye.) (*b*) Here follow two rays (*solid lines*) that diverge from the head of the arrow, point *A*. After passing through the lens the two rays are still diverging. If we extend these refracted rays backward (*the dotted lines*), they meet at point *A'*. Similarly two rays that begin at point *B* (*dashed lines*) are traced through the lens and extended backward to point *B'*. Thus on the right of the lens it appears that rays are diverging from an arrow in the position *B'* to *A'*. That is a virtual image. Not only is there no material body there, but there are also no light rays there. Still, the viewer on the right "sees" a magnified upright arrow. Note that since no light passes in the region of the virtual image, any kind of opaque barrier (a table) can be placed immediately behind the object, and it will not prevent the image from being seen.

(fig. 2.4a). This image is inverted: object points above the optical axis are focused below the axis beyond the lens; object points below the axis are focused above the axis in the image.

The focal point is the place where rays parallel to the axis are focused. Since rays entering the lens from an extremely distant object are almost parallel, we can say that the focal point is the image point of an object at infinity and on the optical axis. The plane parallel to the lens and through the focal point is called the *focal plane*. Since rays of light can pass through the lens in either direction, one can think of two focal points, one on each side of the lens, and two focal planes. The distance from the lens to each focal point is called the *focal length*. This length is the principal characteristic of the lens, and it is determined by the shape of the lens and also by the index of refraction of the glass. Greater curvature of the lens surface and higher index of refraction lead to more bending of rays and hence to a shorter focal length.

When you project the lampshade's image on the wall, you have not, strictly speaking, seen the image, but rather you have seen the reflection of that image from the wall. But if the wall were not there, the rays of light which converge to form the image would not stop; they would continue beyond that plane, now diverging, in the same way that rays of light diverge in all directions from an illuminated object. Thus, if you place your eye beyond

Fig. 2.5. Types of images with a convex lens. (*a*) If the object is slightly outside the focal plane, the image is real, inverted, and magnified. (*b*) If the object is slightly inside the focal plane, the image is virtual, erect, and magnified. (*c*) If the object is far outside the focal plane, the image is real, inverted, and diminished.

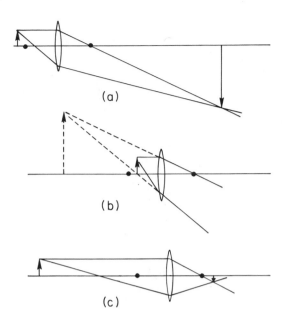

the image and allow the diverging rays to enter the eye, you will see the image. Hold the lens at arm's length and look through it at the lamp across the room; you see, not the lamp, but its real image.

Neither of these experiments illustrates the intended use of the lens, which is to magnify. Consider what happens if an object is placed very close to the lens, so that rays from the object are diverging much more than in the previous illustration. Now the lens cannot converge the rays but can only decrease the degree of their divergence. The eye receives these diverging rays from the lens and interprets them as coming from a more distant and higher point than the actual point on the object. On paper we can trace the rays back to a hypothetical point from which they appear to have come, on the object side of the lens. This point (or infinite series of points corresponding to the infinite array of points on the entire object) is called a *virtual* image (fig. 2.4b); it is virtual in the sense that although no light rays actually pass through the point, the eye receives rays *as if* an object were there. The image is larger than the object; it is magnified. (see fig. 2.5).

The virtual image is also more distant than the object, and this effect is in a sense the real function of a magnifying glass. For why in fact does one need any help in seeing fine print or any small detail? The eye's retina (as we will see in more detail later) is not a continuous screen but rather an array of receptors of finite size upon which the eye projects a small image of objects seen. If the object is very small, its image on the retina may be so small that it stimulates only one or a few receptors. The observer will then not be able to distinguish any fine detail of that object (for example, a printed letter in a book seen at a distance of 20 feet). The natural strategy would be to move the object much closer to the eye, enlarging the image on the retina. But a very small object may have to be brought very close to the eye, say a few inches, in order to make its image on the retina large enough; and here one reaches a limitation on the eye's power of *accommodation*, or bringing to a focus objects at different distances. The eye cannot focus on objects closer than about 10 inches. (This distance varies among individuals; in particular, it increases with age.) The magnifying glass, therefore, allows the eye to look, not at the object, but at a more distant virtual image, which the eye *can* focus; yet the image is larger than the object, so the image produced on the retina is magnified. Try using a magnifier this way: Hold it right against the eye, with the eye 5 or 6 inches above a page of print. You should see the print clearly and be able to distinguish the fine detail of the letters. Now, without moving your head, remove the lens. The letters look the same size, indicating that the image on the retina is the same size, but you cannot bring them into focus.

Optical Instruments

Using a magnifying glass in combination with a second lens, we construct a compound microscope (fig. 2.6). At one end of a tube place a convex lens, called the *objective lens*, and place the object to be viewed just outside the lens's focal point. A magnified real image is produced on the viewer's side of the lens. At the other end of the tube place a second lens, called the *eyepiece*, so that the real image is located between the two lenses, close to the eyepiece. The eyepiece is now a magnifying glass, treating this real image as if it were an object, producing from *it* a further magnified virtual image. This final image is located sufficiently far behind the eyepiece that the eye, held against the eyepiece, can view it comfortably.[2] The total magnification is the product of the magnifications of the two separate lenses. Typically the objective magnification can be from 10 to 100; the eyepiece magnification from 5 to 10; the total magnification from 50 to 1,000, or sometimes up to 2,000.

The inventors of the microscope (Zacharias Janssen) and of the telescope (Hans Lippershey) were both Dutchmen, working around the year 1600. Indeed in the sixteenth and seventeenth centuries the Netherlands was the center of the production of lenses for spectacles. So it is not surprising that these two major advancements in the tools of science occurred at about the same place and time. The great seventeenth-century philosopher Baruch Spinoza spent much of his life grinding and polishing lenses for microscopes, telescopes, and spectacles in the towns of the Netherlands.

Like the microscope, a telescope (fig. 2.7) may be made of two convex lenses, one objective, directed at the object being viewed, and one eyepiece held up against the eye. The difference between the two instruments lies mainly in the location of the object, for with a telescope we view an object (the moon, for example) that is at a great distance and is itself very large. Thus the objective does not magnify in the sense of producing an image larger than the object. (We have no desire to produce an image larger than the moon!) It is useful to think not so much of the object itself but of a series of nearly parallel light rays arriving from a point on the object (say, the upper edge of the moon) and another series of light rays arriving from a different point (say, the lower edge of the moon). Between these two series of lines there is an angle (about 0.5 degrees in the case of the moon), and this angle determines what we call the *angular size* of the object. The function of the telescope is to produce an image with a larger angular size than the object.

Each series of parallel rays passing through the objective converges to a point behind the lens, forming an inverted image of the moon in the focal plane (just as earlier we projected an image of a lampshade on a wall). The eyepiece is placed beyond the

Fig. 2.6. Principle of the compound microscope. The objective lens forms a real image (I_1) just inside the focal point of the eyepiece. The eyepiece then produces a magnified virtual image (I_2) of I_1. (For the sake of clarity, the magnification shown here is less than what it would be in practice.)

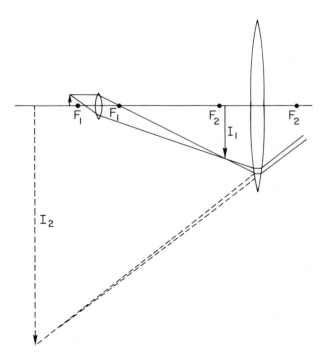

28

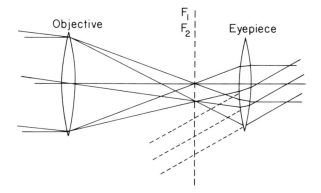

Objective F_1 F_2 Eyepiece

Fig. 2.7. Principle of the telescope. Rays parallel to the optical axis are brought to a focus at the focal point. Rays which are parallel to each other and which make a certain incoming angle with the axis are brought to a focus in the focal plane, F_1, but below the axis. This plane coincides with F_2, the focal plane of the eyepiece. Beyond this plane the rays are diverging. They enter the eyepiece and emerge in such a way that rays which were parallel when they entered the telescope are again parallel. But the angle between the off-axis rays and the optical axis is now much larger than the incoming angle.

image, such that the image falls precisely in its focal plane. The optical effect of this second lens will be just the opposite of the effect of the objective. Where the objective converges parallel rays to a point in the focal plane, the eyepiece receives rays that diverge from a point in *its* focal plane. Therefore its effect is the reverse: it produces from these diverging rays a series of parallel rays. The key to magnification in a telescope is that the eyepiece is a short focal length lens: it bends light rays much more effectively than the objective. So two series of rays that entered the telescope with an angle of 0.5 degrees will emerge from the eyepiece with a greatly increased angle, say 10 or 40 degrees, depending on the lenses. The magnification is just equal to the ratio of the focal lengths.

    The magnification of a telescope, therefore, is magnification of the angular size of an object, and it is important to realize that the concept of angular size is a bridge between optics, the behavior of light passing through a lens system, and perception, the interpretation by a person's neural system of light signals received through the eye. Light rays enter the eye at a certain angle, to be focused upon the retina, and the only visual information the brain receives about the outside world is just that, the *directions* of rays. Therefore, it is the angular size of an object that determines the apparent size of an object in one's consciousness. Two objects with the same angular size will appear to have the same size. A good example is the sun and the moon, which look about the same size in the sky, although the sun is about 400 times larger and correspondingly 400 times farther away. Every child discovers this paradox of distance and size. With her arm outstretched she holds up her thumb and sees resting on top of it an automobile at the end of the road. Is it a toy car balanced on her thumb or a real automobile at a distance? Only secondary cues—the house next to the car, the haze of the air above the road—give her the answer.

    There is a limit to how short the focal length of a precision

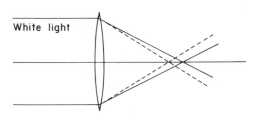

White light

Fig. 2.8. Chromatic
aberration. Red light
(*solid lines*) is
focused at a farther
point than violet light
(*dashed lines*).

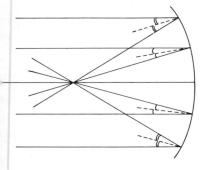

Fig. 2.9. Focusing with a
curved mirror. The dashed
lines are drawn normal (perpendicular) to the mirror surface. Each incident ray is
reflected to an equal angle on
the opposite side of the normal
(i.e., the angle of reflection
equals the angle of incidence).

eyepiece can be made, and this places a limitation on the magnifying power of a telescope. But for many uses, in fact for the most
important uses of a telescope, magnification is not important. Indeed, the eyepiece is dispensed with altogether, and the image is
projected directly on a photographic plate. When a telescope is
used to probe the outer reaches of space, to study stars and distant
galaxies, its main function is to collect a large amount of light.
The objective lens is made as large as possible; it then collects
many parallel rays from a star and projects them all onto a point in
the focal plane where the photographic emulsion lies. Thus it
allows us to "see" a distant object—a star—which may be so
faint that it is completely invisible when viewed through an ordinary camera lens or through the narrow pupil of the eye. (In addition, photography allows the astronomer to collect light over long
"exposure" times, often many hours, and thus detect sources that
would be otherwise invisible.)

The laws of lens focusing are complicated by the fact, as
Newton had discovered, that different colors of the spectrum are
bent differently by glass. Thus all lenses suffer from *chromatic
aberration* (fig. 2.8), the focusing of different colors at slightly
different points along the axis. Newton was therefore led to construct a telescope with a concave mirror as objective, instead of a
lens (fig. 2.9). Because the law of reflection is independent of
wavelength, the real image produced in front of the mirror is
formed at the same point for all colors. Today large telescopes are
always made with mirror objectives (figs. 2.10, 2.11, and 2.12).

In the case of the microscope, the problem of chromatic aberration is solved by using as an objective what is called an
achromatic lens, one made by cementing together two lenses
made of different types of glass, each with a different index of
refraction. The combined system is convergent, and the curvatures of the two glass-air boundaries may be chosen so that the
focal point is the same for *two* colors, usually red and blue. The
focus is then not entirely independent of color, but its variation
over the spectrum becomes very small. The development of the
achromatic microscope in the nineteenth century was in fact a
major step in the advance of science. It permitted observations on
a much smaller scale than previously, leading to important prog-

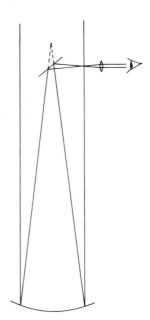

Fig. 2.10. Exterior view of the Mayall 4-meter telescope building (Courtesy National Optical Astronomy Observatories)

Fig. 2.12. Interior view of the 41-inch reflecting telescope at the Yerkes Observatory (Courtesy Yerkes Observatory)

Fig. 2.11. The reflecting telescope. A real image is formed in front of a concave mirror. The image is therefore in the path of the incoming light. To view it with the eye or a camera one must block some of this incoming light. Newton placed a small flat mirror at a 45-degree angle to the axis, shortly before the focal point. This mirror reflects the light toward the side of the telescope, where the image can be viewed out of the light path. (The small mirror does not create a hole in the picture. It only slightly diminishes the intensity of the entire image.) In modern telescopes more complicated mirror systems are used to bring the image to a convenient location for further analysis, which may involve a camera, a spectroscope, or a photoelectric cell.

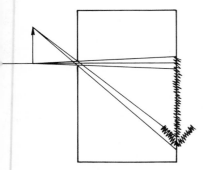

Fig. 2.13. The pinhole camera. Rays from a given object point form a fuzzy spot of light inside the camera. The smaller the pinhole, the sharper the image. But a small hole requires a long exposure time to supply enough light to expose the film.

ress in bacteriology and the treatment of such diseases as anthrax, cholera, tuberculosis, and yellow fever.

Even with monochromatic light, however, a lens or mirror made in the traditional way—by grinding glass or metal surfaces into the shape of a portion of the sphere—suffers from other "aberrations"; it fails to produce a perfect image. Parallel rays passing through the lens (or reflected from a mirror) do not quite meet at a precise point; the image in the focal plane is a small blur. Large mirrors used in telescopes are therefore ground in the shape of a paraboloid, that is, a parabola rotated about the telescope axis; this curvature focuses parallel rays at a precise point. Smaller optical systems use a combination of several lenses, sometimes cemented (as in the achromatic lens), sometimes separated by an air space, to minimize aberrations. But a practical optical system is not perfect and represents a compromise among various desired properties, including sharpness of focus and width of the field of vision.

An interesting example of an optical system is the camera, in which an image is projected onto film by a lens or a series of lenses. In principle a camera can form an image without a lens, with just a small hole into a light-sealed box. If we trace light rays from a point on an outside object through the hole, we find that only a narrow cone of light enters the box; each object point is imaged as a fairly small exposed region in the interior (fig. 2.13). To make the image sharp the hole must be very small, and this means the film must be exposed for a fairly long time, to admit enough light to affect the film. Thus the pinhole camera is adequate only for photographing still objects. If the object is moving, or, more significantly, if the camera is hand-held and so subject to involuntary motions made by the photographer, then the exposure must be made in a short time—typically one-thirtieth of a second but sometimes shorter than one-thousandth of a second. In order for there to be enough light, the camera opening (its aperture) must be made larger. As a consequence, the camera must have a lens to focus light. The large aperture, however, leads to aberrations. These in turn are corrected for by introducing a more complex system of lenses.

The art and science of lens-making involves fashioning lens surfaces and combinations of lenses in such a way that light rays from an object point, traced through the system, meet as nearly as possible in an image point. A perfectly ground parabolic mirror in a telescope will indeed focus parallel rays to a point. But two further effects make it impossible, even under these conditions, to produce a perfect image. In practice the more important effect has to do with turbulence in the upper atmosphere. Light rays from, say, a distant star are refracted through rapidly moving air and so are not all parallel when they reach they earth's surface. The eye perceives this as the twinkling of stars. On film the result is a fuzzy image in a roughly circular region. When two objects are close together in the sky (i.e., have a small angular separation), their fuzzy images may so far overlap that we cannot distinguish two lit regions on film; we say the two objects cannot be *resolved*. Thus turbulence limits the ability of ground-based astronomers to resolve objects, such as binary stars, or to see fine details, such as small lunar craters. A telescope's resolving power may be characterized by the minimum angular distance in the sky between two objects that can be distinguished.

To avoid the problem of atmospheric turbulence, astronomers have planned the earth-orbiting Space Telescope, originally scheduled for launch by the space shuttle in 1986. (It will also be able to study electromagnetic radiation over a wider wavelength range than is possible with ground-based telescopes, because much of the infrared and ultraviolet from space does not get through the atmosphere at all.) Free of atmospheric limitations, a lens or mirror comes upon a more fundamental limit on focusing, namely, the wave character of light. The treatment of light propagation along a ray (like the trajectory of a baseball) is only an approximation. A light wave bends around corners, creating diffraction patterns at the edge of a shadow; and in place of a point of light in the focal plane there will be a central bright spot and fainter concentric rings. The diffraction effect is small, because light wavelengths are small, but in precise optical instruments it is not negligible. It is the final limit on a telescope's resolving power. The minimum angle of two objects that can be resolved varies directly with the wavelength and inversely with the diameter of the objective. As an illustration, an amateur telescope with a 6-inch mirror, receiving light of wavelength 500 nm, has a resolution angle of $1/5,000$ of a degree. This may seem like a comfortably small angle. But astronomers are concerned with objects at extremely great distances that have correspondingly small angular sizes. If the sun were placed at the distance of the nearest star, its angular size would be reduced from $1/2$ of a degree to $1/500,000$ of a degree. So the 6-inch telescope, indeed even a 200-inch telescope, could not resolve such a star. It would detect the star but could not determine its size, shape, or any feature upon it. The diffraction problem becomes still more crucial when we look at radiation of much longer wavelength, as in radio astronomy. It is not surprising therefore that astronomers have built bigger and bigger objectives, not only to increase the radiation-gathering power but also to improve the resolving power of telescopes.

The wave properties of light also limit the effectiveness of the microscope. For light reflected from—or transmitted through—an object smaller, roughly, than the wavelength of that light will not produce a pattern representing the shape and structure of the

Fig. 2.14. Diffuse and specular reflection. (*top*) Diffuse reflection occurs when light is reflected from a rough surface. The direction of the reflected ray is determined by the same law of reflection at each surface point. But since the surface is rough, reflected rays return in many different directions. (*bottom*) Specular reflection occurs when light is reflected from a smooth surface. All reflected rays return in the same direction.

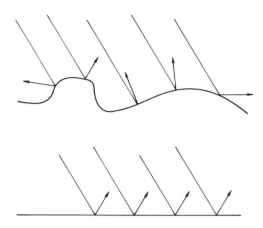

object but rather will produce a diffraction pattern characteristic of the light wave. Fortunately typical sizes of biological cells, including one-celled organisms, are a thousand or several thousand nanometers, dimensions greater than the wavelengths of visible light. Hence the light microscope was able to open the way to modern biology.

The wave properties of light also determine what kind of surface can act as a mirror. Almost every surface reflects light: the page of a book, a painted wall, the nose on your face. But these surfaces are not mirrors, because they are not optically smooth—that is, they vary up and down over a distance scale larger than the wavelengths of light. The laws of reflection do apply to light striking a rough surface, but at each point of the surface the normal to that surface is in a different direction, and these normals vary in a somewhat random way across the surface (fig. 2.14). Therefore light is reflected from the surface in all directions, bearing no relation to the direction of the incoming beam. We have what is called *diffuse* reflection: the surface can be seen, but no image of the object producing the beam is formed.[3] If, on the other hand, the surface imperfections are smaller than the wavelength, then the wave is not affected by them, and it is reflected on the other side of the normal, in a direction determined by the incoming beam. This is *specular* reflection. Diverging rays from an object point are then reflected from the mirror in such a way that they appear to be diverging again from a point behind the mirror. A virtual image is created at a distance behind the mirror equal to the distance from the object to the mirror (fig. 2.15).

The Eye

When we turn from precision optical instruments to the subject of the human eye we confront the profound distinction between evolution and engineering. The engineer is asked to design an instrument to carry out a given function, to meet certain specifications. He is given his materials—glass, metal, etc.—and attempts to find the optimum design for his instrument. Evolution, on the other hand, is the modification of an existing organism, making use of parts and functions already in place. Genetic changes that are good or useful are then passed on to succeeding generations. Everywhere in the natural world we see examples of such adaptation: the bat's wing for flying, the seal's flippers for swimming.

Among the vertebrates the basic structure of the eye is similar, although we see an endless variety of adaptations to the different habits and environments of species. The human eye is typical, being neither primitive nor highly specialized. Indeed because of our reliance on intelligence, most of our physical attributes are not extraordinary, our vision surpassed by the hawk's, our nose surpassed by the hound's.

The two most important parts of the vertebrate eye (fig. 2.16) are at the front and the back: the cornea to focus an image and the retina to record the image. The lens, behind the cornea, plays a subsidiary role in focusing; by changing its shape it carries out minor modifications in the refraction of light to allow us to focus on near objects. The iris, resting upon the front of the lens, expands and contracts to regulate the size of the central opening, the pupil. The pupil's size determines how much light passes into the eye. Between the lens and the cornea is a fluid, the *aqueous humor;* behind the lens, filling the cavity surrounded by the retina, is a more rigid, gelatinous substance, the *vitreous* (meaning

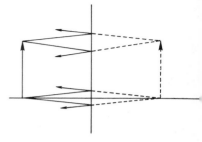

Fig. 2.15. A virtual image in a plane mirror. Rays from a given object point are reflected from the mirror and diverge as they return. If we trace the rays backward (*dashed lines*), we find they meet at a point just as far behind the mirror as the object is in front of it. If you stand 5 feet in front of the mirror, your image is 10 feet away from you. If the surface were rough, the directions of the reflected rays would be changed in a more or less random way and they would not all meet at an image point.

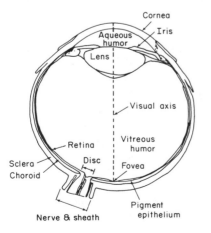

Fig. 2.16. A cross section of the human eye. The *sclera* is an opaque continuation of the cornea. The *choroid* coat is a network of blood vessels behind the pigment epithelium and the retina. (Courtesy Cranbrook Institute of Science)

"glassy") *humor.* Around the back of the eyeball is spread the retina. Its light-sensitive visual cells, or *photoreceptors,* are located—perhaps surprisingly—in its outer layer (next to the pigment epithelium), away from the source of light. The inner layers of the retina are transparent and contain nerve cells which receive signals from the visual cells. These nerve cells eventually join to form the optic nerve, which passes through the retina, out behind the eye, and to the brain.

The visual cells that form the outer layer of the retina are of two types: rods and cones (see fig. 2.20). The system of rods is highly sensitive and useful in dim light; the cones are comparatively less sensitive but provide high acuity and color vision. The light-sensitive portion of these cells is called the *outer segment* (nearest the back of the retina; see fig. 2.23). In the rod of the human eye the outer segment is a cylinder about 0.002 cm long and 0.0002 cm in diameter. The cone outer segment is shorter and tapers in some species (hence the name).

The eye is an effective optical instrument because its flexible casing is kept firm by the pressure of the fluids within. Thus the curvatures of its refractive surfaces can be held constant. The vitreous humor is more or less permanent, while the aqueous humor is continually drained and replaced by new fluid. (Failure of the drainage system leads to excess pressure in the eye, the condition known as glaucoma.)

Because the wavelength of light is small, it is fair to picture the travel of a light wave as a series of rays that bend at each interface, each ray entering a single photoreceptor. The effects of diffraction, in other words, are not extremely important in the eye. But the photoreceptor is also small, its diameter only about four times the wavelength of light. If visible wavelengths were at all larger, we would begin to lose acuity because of diffraction; that is, the diffraction pattern of a point of light would overlap several receptors in the retina.

The cornea evolved from the skin of the head; it merges with the inner layer of the eyelid and hence is continuous with the skin. The cornea is curved approximately like a section of a sphere. Its index of refraction is 1.376, and so light entering the eye is refracted and made to converge by the air-cornea interface. Note that this is not a lens of the type discussed earlier in the chapter, which consisted of two interfaces, air-glass and glass-air, of opposite curvatures. After light enters the cornea it passes through the aqueous humor (index of refraction 1.336), the lens (1.42), and the vitreous humor (1.336). Refraction at these farther surfaces is relatively slight (because the speed of light does not change much), but the lens is capable of some additional converging power.

When the lens is relaxed, an image formed of a very distant object will lie in the forward ends of the outer segments of the visual cells (along the inner edge of the retina's photosensitive layer). If the object moves closer and the optical system is unchanged, the image moves backward; but it may still lie in the photosensitive layer (which has a thickness of 0.002 cm, the length of the outer segments). The human eye can in this way see a focused image for an object at any distance between infinity and about 20 feet.

To bring nearer objects into focus on the retina the eye must accommodate, and this is the function of the lens. Having a greater index of refraction than the humors around it, the lens contributes to convergence of the light. In man, and mammals generally, the lens accommodates by means of its own elasticity: fibers that ordinarily hold the lens in a somewhat flattened shape are relaxed, allowing it to increase its curvature and thus its converging power. Birds and most reptiles change the lens shape by means of muscles that compress its perimeter. Some animals, including fish and amphibians, accommodate not by changing the shape of the lens but by moving it forward or back. Not all species have to accommodate for near objects. For certain fish and lampreys, near objects are in focus when the eye is relaxed, and the lens must move backward for distant objects. Those reptiles that catch fast-moving insects for a living have the fastest accommodation among the vertebrates. Fish are the earliest major vertebrate form, so it is not surprising to find relics of the fish eye in later species. For example, the aqueous humor is liquid, probably to allow for forward-backward motion of the lens, a process that no longer occurs in man.

There is a limit to how close an object can be brought to the eye and still be accommodated by the lens. In humans this close point is roughly 10 inches; it increases significantly with age because the lens hardens. Past the age of forty the close point usually increases well beyond 10 inches. This is normal farsightedness, in which one cannot read without holding the printing far from the eye (fig. 2.17). It is corrected by the use of convex lenses, which provide additional converging power.

Nearsightedness exists when the eye is too deep, so that an object at infinity is focused not on the retina but too far forward, in the vitreous humor. A closer object will be focused farther back in the eye, so the nearsighted person may see close objects in focus without accommodation. Nearsightedness is corrected by diverging, or concave, lenses.

The third common malfunction of the human eye is astigmatism. This occurs when the shape of the cornea is such that it has a different rate of curvature along one direction (say, a hori-

Fig. 2.17. (*top*) Farsightedness. The lens cannot accommodate near objects. The point
at which they would be
focused lies behind the retina.
(*bottom*) Nearsightedness. Distant objects are not focused on
the retina. With the lens relaxed, the eye is too deep, and
parallel rays are focused in
front of the retina.

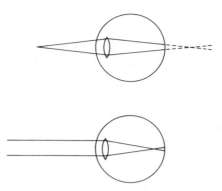

zontal circle) than along another (a vertical circle). The result is
that a point in the visual field is imaged as an elongated spot on
the retina. Astigmatism may be corrected by means of an eyeglass
lens which itself is astigmatic (has different rates of curvature) in
a way opposite to that of the cornea. Such lenses must be correctly oriented over the eye and prevented from rotating about an
axis through the eye.

One malfunction that cannot be corrected is chromatic aberration. The eye's optical system focuses different colors in different focal planes—blue most forward (toward the front of the eye),
red most backward—and the difference between the two focal
planes is greater than the thickness of the photosensitive layer of
the retina. If you view a series of red and green stripes and try to
focus sharply on the red, the green will be slightly out of focus.
The brain will interpret this as a different depth; the red stripes
seem to be closer.

To return briefly to the water, sea-dwelling creatures face an
optical problem not encountered by those who live in air: the index of refraction of water is practically the same as that of the
cornea and the humors. Hence light is refracted relatively little on
entering the eye, and most of the work of focusing falls to the
lens. The index of refraction of the fish lens thus tends to be
higher than those of land animal lenses. Moreover the fish lens is a
sphere, providing as much curvature as possible for greater converging power. As one passes from fish to amphibious animals to
land animals, in the direction of evolutionary history, there is a
gradual flattening of the lens.

Amphibious creatures face a formidable problem if they
wish to see both in air and in water, for the cornea is optically
more effective in one medium than in the other. If a man opens his
eyes under water he is extremely farsighted; that is, just as a farsighted person cannot focus on an object held close to his eye, the
swimmer cannot focus on an object at any distance. Many amphibious species actually do most of their seeing in air (frogs and

crocodiles, for example) and simply have poor vision in water. Turtles have a particularly soft lens acted on by unusually strong muscles, so that they can accommodate for both air and water vision. Certain diving birds have evolved a special use for their nictitating membranes, a "third eyelid," possessed by birds and reptiles. This transparent membrane closes sideways across the eye and in general serves the purpose of cleaning and moistening the eye without the animal's being temporarily blinded. In diving ducks and loons the nictitating membrane has a central region with a high index of refraction, which is brought into play to provide focusing under water.

An exotic solution to the air-water problem is found in the cuatro ojos, a "four-eyed" fish found in Central and South America (fig. 2.18). It swims along the water surface with its head partly out of water, and each of its two eyes has two pupils, one kept above the surface, the other below. Each pupil focuses upon a different retina. But there is one lens, of elliptical shape, cleverly oriented so that the pupil in water focuses through the more sharply curved surfaces of the lens, while the pupil in air focuses through the less sharply curved surfaces. The fish thus obtains two focused images, one of the sky and one of the sea. Still odder is the eye of *Dialommus fuscus,* a small amphibious fish of the Galapagos Islands. Its eye has one pupil divided horizontally into two plane areas, and the cornea is in the form of a prism. This fish receives a double image of any object in the forward direction. The adaptive value of this peculiarity is not certain.

In addition to seeing near and far, the eye must also adapt itself to high and low intensities of light. Most nocturnal species have adapted by evolving eyes with more rods than cones; diurnal species have more cones. Lizards and most snakes have only cones; birds have more cones than rods. The human retina has about seven million cones, but twenty times as many rods. This appears to be a large rod-cone ratio for an animal that spends so many hours in the sun. But in comparing species we must remember that the size of the visual cells is approximately the same for all species. Therefore a larger animal has room for more visual cells altogether. A human eye can obtain a sharp image with seven million cones and still has room in the retina for many more rods; a small bird fills its retina with several million cones. Fish have both rods and cones, but rods predominate in species that frequent deep water; some deepwater fish have only rods. The rod-cone ratio has another significance beyond sensitivity, which we will come to shortly.

A further word, however, is in order about vision in the sea. Light is strongly absorbed by water. The extent of absorption varies with the nature of the water, but, roughly, 90 percent of the light of the atmosphere is lost at a depth of 30 feet. Below a few

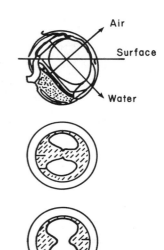

Fig. 2.18. The eye of cuatro ojos. (*top*) Vertical cross section. The eye has two pupils, one looking into the water, the other into the air. The former uses the more sharply curved edge of the elliptical lens to help converge light on the retina. The latter pupil uses the less sharply curved edge of the lens. (*middle*) Front view of the double pupil. (*bottom*) Pupil of the animal in the larval stage, showing that the double pupil develops from a single one. (Courtesy Cranbrook Institute of Science)

Fig. 2.19. The round pupil (*top*) and the slit pupil (*bottom*). The round pupil closes by means of a circular sphincter muscle. The slit pupil also has a sphincter, but it is aided in closing by contraction of additional vertical fibers. Both pupils dilate by means of radial muscles. (Courtesy Cranbrook Institute of Science)

hundred yards there is not enough skylight to affect a retina. Yet deep-sea fish are not blind. In fact, as depth increases the eyes of fish at first become larger, increasing sensitivity. Only at the greatest depths do the eyes diminish again, for vision becomes impossible or less important. Some species found in very deep water or in caves are blind, although, interestingly, they are not eyeless. Rather the eyes have degenerated or have been covered with an opaque layer, indicating that sight was available to the ancestors of these creatures and that the sense gradually withered in those species that sought a dark environment. But why do we find vision at all in the deep where sunlight never penetrates? The answer seems to be that many deep-sea fish (two-thirds of all species) are bioluminescent; that is, they produce light from special organs along their sides (see chap. 8). Thus a fish sees other fish, his enemy, or his dinner.

Beyond these more-or-less permanent adaptations for dark and bright habitats, an animal must be prepared to adjust to changing levels of illumination. The contractile pupil is one device for this purpose. When illumination decreases, the iris of the human eye is opened involuntarily by a series of radial muscles that dilate the pupil and admit more light. The pupil is contracted by a sphincter muscle. For some animals—those with nocturnal habits and sensitive retinas—closing of the pupil in bright light is very important. Thus, many mammalian carnivores and reptiles and some primates have evolved the slit pupil, which is capable of a greater degree of closure by contraction of muscles along the sides of the pupil (fig. 2.19). In only a few species, such as the smaller cats, can the slit be completely closed.

A second technique for adjusting to varying brightness is used by many fishes and birds. Behind the retina of the vertebrate eye there is a layer of heavily pigmented cells, called the *pigment epithelium*. This layer normally absorbs much of the light that fails to interact with the photoreceptors. In some animals the cells of the pigment epithelium are mobile, and in response to bright light they travel forward and surround the outer segments of the visual cells. They absorb light that might enter through the sides, or be scattered from behind the retina. This device is prevalent lower on the phylogenetic scale, and is absent in mammals. It acts slowly, and is suitable largely for adaptation at dawn and dusk.

The Visual Cells

These are the mechanisms that act in response to light as it passes through the eye's optical system toward the photoreceptor. Yet there must be more to light-dark adaptation, for we are able to see over intensity levels varying by a factor of more than 10^{12}, a mil-

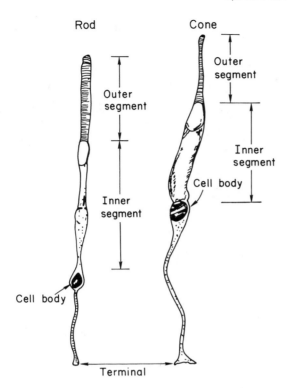

Rod

Cone

Outer segment

Outer segment

Inner segment

Inner segment

Cell body

Cell body

Terminal

Fig. 2.20. Structure of rods and cones. The cell body is the region where the nucleus is found. (Adapted from Michael Levin and Jeremy M. Shefner, *Fundamentals of Sensation and Perception* [New York: Random House, 1981], 91.)

lion million, while the area of the human pupil can change by not more than a factor of about 16. It must be that beyond the photoreceptors, after light has been absorbed, there are neural systems that amplify weak signals and enable us to respond both in starlight and in daylight. Before venturing into these regions, let us pause to examine the photoreceptor in detail (fig. 2.20), since it is the central point in vision: the end of the optical system and the beginning of the neural system.

The outer segment of the visual cell contains a light-absorbing pigment and points backward, away from the light source. Further forward is the *inner segment,* a transparent cylindrical body that funnels light along its axis toward the pigmented outer segment. Experiments have shown that it acts as a light pipe, like the fiber-optic systems used by doctors to see into the body (see chap. 9).

Seen in cross section under an electron microscope (fig. 2.21), the outer segment consists of closely packed layers of a membrane in which is embedded the visual pigment. *Rhodopsin,* the pigment of the rods, is deep red, showing that it absorbs all wavelengths of light except for the longest. It is possible to measure the *absorption spectrum* of an individual rod—that

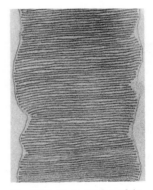

Fig. 2.21. Cross section of the outer segment of the rod of a rat, viewed in the electron microscope. Magnification is approximately 100,000. (Courtesy John E. Dowling; reprinted, by permission, from John M. Allen, ed., *Molecular Organization and Biological Function* [New York: Harper and Row, 1967], 192.)

Fig. 2.22. Schematic diagram
of the structure of the primate
retina, as revealed in the electron microscope. (Adapted
from J. E. Dowling and B. B.
Boycott, *Proceedings of the
Royal Society of London* 166B
[1969]: 104.)

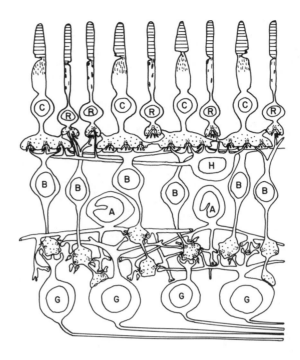

is, the curve showing its absorption of light as a function of wavelength—by using a photocell to detect the light transmitted through the rod.[4] When rhodopsin is subjected to intense light, however, it is bleached; it becomes transparent, losing much of its absorptive capacity. When the sensitivity of the rod system is destroyed, we see with the less-sensitive cone system. In the dark again rhodopsin will automatically regenerate, in a period of about 30 minutes.

A single rod is only slightly more sensitive than a single cone. The major differences between rods and cones (apart from color sensitivity, which we take up in the next chapter) are their distribution across the retina and the way in which they are connected to neural transmitters. Inside the layer of visual cells lies a layer of what are called *bipolar* cells (see figs. 2.22 and 2.23). The neural connections between these two layers are not one-to-one; rather, many visual cells may be connected to one bipolar cell. Bipolar cells are in turn connected to nerve cells in the next layer, the layer of *ganglion* cells. Again several bipolar cells converge on one ganglion cell. From the inner ends of the ganglion cells nerve fibers emerge and traverse the inner boundary of the retina until they meet to form the optic nerve. As a result of this convergence, although there are about 150 million photoreceptors in the retina, there are only about one million fibers in the optic nerve. Furthermore, convergence occurs to a much greater extent for rods than for cones, and in this way the rod system is made highly sensitive (several rods are more likely to excite a response in a single nerve cell than is one rod); rods become our window on dim light. But there is another consequence of convergence: a loss of visual acuity.

Remember that the image on the retina is not a continuous distribution of light and dark but is, in effect, a large mosaic of light and dark spots. Each visual cell either absorbs light or does not. This mosaic structure places a limitation upon how sharp an image we receive, that is, upon our acuity. For example, two spots of light so close together that their projections on the retina fall on the same visual cell cannot be perceived as two separate objects. If many visual cells trigger a single optic nerve fiber, we lose acuity. In dim light our vision is poor, not *directly* because of the lack of light but because our cones are inoperative. We rely on rod vision, which has low acuity as a result of extensive convergence.[5]

In the animal kingdom nature makes a choice or compromise between sensitivity and acuity. For creatures that fly acuity is essential; they must give up sensitivity: birds have primarily cones and must sleep at night. The cat and the opossum work at night and so settle for an acuity much poorer than that of humans.

When an animal has both rods and cones in the retina, there is frequently a central area, the *macula,* where the cones are con-

Fig. 2.23. Microscope photograph of a human retina. Magnification is approximately 400. (Courtesy John E. Dowling; reprinted, by permission, from John M. Allen, ed., *Molecular Organization and Biological Function* [New York: Harper and Row, 1967], 188.)

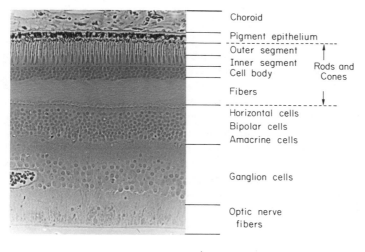

Choroid
Pigment epithelium
Outer segment
Inner segment
Cell body — Rods and Cones
Fibers
Horizontal cells
Bipolar cells
Amacrine cells
Ganglion cells
Optic nerve fibers

LIGHT

Fig. 2.24. Distribution of rods
and cones across the retina.
The number of visual cells per
unit area is plotted against the
angular position, as defined in
the upper diagram. (Redrawn
from Tom N. Cornsweet, *Visual Perception* [New York:
Academic Press, 1970], 137;
and M. H. Pirenne, *Vision and
the Eye* [London: Chapman
and Hall, 1967], 32.)

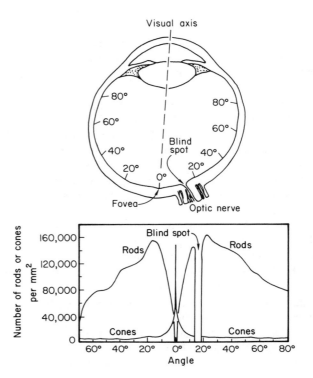

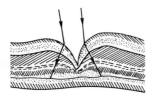

Fig. 2.25. Fovea of the European bank swallow. Light rays
are spread apart by a deep
fovea, magnifying the image
on the layer of visual cells.
(Courtesy Cranbrook Institute
of Science)

centrated (fig. 2.24). The human looking intensely at a small object is directing the image onto this central area, where vision is sharpest. On the other hand, one cannot use this area at night, since it has no rods. You may have had the experience of catching a dim star in your peripheral vision, but when you try to fixate on it, it disappears. Your cones do not work in the dim light, but your rods, off center, do.

In the macula of the human eye there is a small hollow or depression, the *fovea*. Its function, if any, is not well understood. But birds have a deeper depression, which presents two sharply curved surfaces to the incoming light. The material of the retina has a greater index of refraction than the vitreous humor, so that light rays approaching the fovea are spread apart, magnifying the image of the visual layers (fig. 2.25). Here is fixed the image of a worm, as a bird swoops down upon it.

How much more intimately can we probe the essence of vision? What is the final step in the transformation of light into a neural signal proceeding toward the brain? Modern structural biochemistry has unraveled this transformation at the molecular level.

A single rhodopsin molecule consists of two parts: a protein molecule, *opsin,* and a relatively simple organic molecule called *retinal*. Like all proteins, opsin is built of many smaller units called amino acids and consists of many thousands of atoms; its complete structure has not been elucidated.[6] Several different proteins are found in cones, and it is the structure of these that determines the response of the photoreceptors to different colors.

Retinal seems to be the unique element in all photopigments—human rods and cones and the visual cells of all animals. Its chemical formula is $C_{20}H_{28}O$: twenty carbon atoms, twenty-eight hydrogen atoms, and one oxygen atom (fig. 2.26). This formula, however, obscures the significance of retinal, for often in organic chemistry two substances may be constructed of the same kinds and numbers of atoms but have very different properties because of different geometrical arrangements of these atoms in the molecule. A change from one geometrical form to another is called isomerization, and it lies at the heart of the function of rhodopsin. In the dark-adapted eye, rhodopsin contains the partially bent structure called 11-*cis* retinal. Upon absorption of light, the molecule is changed to the straight structure, called all-*trans* retinal. In the dark, the 11-*cis* form is attached to the protein opsin, forming the pigmented molecule rhodopsin. The all-*trans* form, in a series of steps, escapes from the protein (fig. 2.27), producing the bleached substance of the light-insensitive rod. Retinal spontaneously returns to the protein, so that in the dark the rod is soon regenerated. Retinal is closely related to vitamin A, the lack of which leads to an insufficient supply of rhodopsin and night blindness.

Vision, of course, is far more than the collection of light in a photoreceptor. It is also the processing of signals from the visual

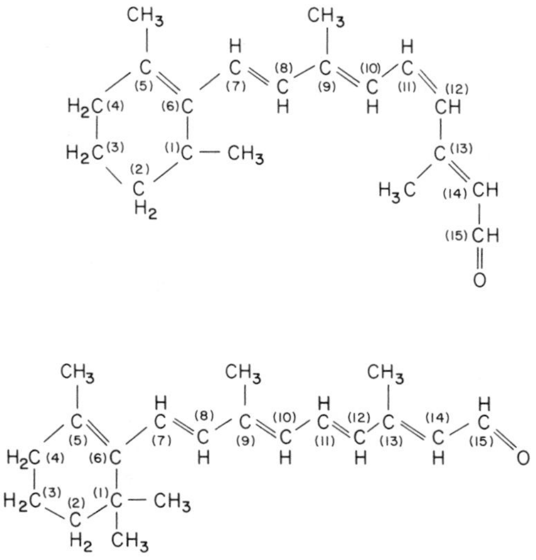

Fig. 2.26. The molecular structure of retinal. In the diagram C, H, and O are carbon, hydrogen, and oxygen atoms, and the short lines represent "bonds" holding these atoms together. Each bond means a sharing of an electron by two adjacent atoms. Some are double bonds, the sharing of two electrons. The 11-*cis* form of retinal (*top*) is the structure when the molecule is part of the protein in the dark-adapted rod. The all-*trans* form (*bottom*) is produced by the absorption of light and causes the retinal molecule to separate from the protein. The numbers in parentheses indicate the number of carbon atoms along the chain. The terms *cis* and *trans* are from Latin, meaning "on the same side" and "on the opposite side," respectively; they refer to the positions of subgroups of atoms with respect to a double bond between carbon atoms. The number 11 refers to the position of a carbon atom along the chain.

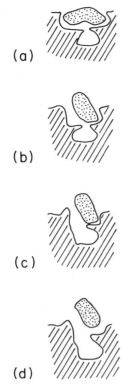

(a)

(b)

(c)

(d)

Fig. 2.27. The states of rod pigment, rhodopsin. Upon absorption of light the molecular structure of retinal changes, and the new straighter form breaks loose at one end from the protein. This is the transition from state (a) to state (b). Subsequent transitions to (c) and (d) occur without the action of light. (Redrawn from G. Wald, P. K. Brown, and I. R. Gibbons, *Journal of the Optical Society of America* 53 [1963]: 22.)

cells at various stages from the retina to the brain and the final creation of a perception in that poorly understood place we call consciousness. In the retina itself, between the visual layer and the ganglion layer, there exist not only the bipolar cells that connect the two layers but other types of nerve cells that make connections in directions parallel to the retina: *horizontal* cells act in the region where rods and cones connect to bipolar cells, and *amacrine* cells act in the region where bipolar cells connect to ganglion cells. Thus data are collected from neighboring parts of the image so that patterns in the image begin to be discerned. The signal carried by the optic nerve to the brain is not simply the two-dimensional mosaic formed in the visual layer but is a message already *processed* by the retina.

The message of a nerve cell is contained in the rate at which the cell "fires," that is, undergoes a brief electric pulse. In the resting state, when unstimulated, the cell fires at irregular intervals, but at a fixed average rate that may be about 50 or 60 pulses per second. When stimulated (whether by touch, sight, or some other influence) the cell responds by either increasing or decreasing its rate of firing. This rate constitutes the signal, and it is transmitted cell by cell along a neural pathway.

In the visual system, the ganglion cells, the cells in the cortex, and others along the pathway between respond in this way. (Photoreceptors and bipolar cells respond differently, by changing a steady electrical voltage across the cell.) A great deal has been learned about nerve cells in the brain and in the retina and how they respond to particular types of stimuli in the visual field.

For example, in the ganglion cells of the cat a single cell may be excited by a spot of light in a given region of the visual field but inhibited by light in the region surrounding that spot. This kind of *excitation-inhibition* response is a widespread phenomenon in vision. In the cat it clearly requires horizontal connections across the retina, tying numbers of neighboring photoreceptors to a single ganglion cell. The excitation-inhibition response enhances contrasts, making an animal more aware of changes in light intensity across its field of view. The cat distinguishes the fly's wing from the noontime sky, and the mechanism for doing so is placed in the eye itself.

In the retinas of rabbits and squirrels certain ganglion cells discriminate motion in a particular direction, responding to the motion of an object, say, to the right but not to the left. Again, neighboring visual cells must cooperate within the retina. The frog's retina possesses cells that respond selectively to certain shapes and motions and to changing illumination. Retinal processing may be more extensive in lower animal forms that have a less-developed brain. (Note, however, the startle reflex in the human when a small moving object is seen "out of the corner of one's eye.")

While our visual system is "programmed" for sensitive detection of structure or patterns in the visual field, our response to uniform light levels over broad areas is relatively weak. But response to the overall light level is important in at least one way. As discussed earlier, we must be able to detect patterns under widely varying conditions of illumination. The light intensity *difference* between an insect and the leaf on which it sits will be very much less at dusk than at noon. The visual system must be able to enhance this contrast at low illumination levels by amplification of neural signals on their way to the brain. This "feedback" mechanism may operate in the retina via the horizontal cells.

We cannot survey the extensive body of experimental data on neural responses along the visual pathways. Cells are found that respond to a variety of complex phenomena (motion, orientation, contrast, color) in the visual field. The eye is remarkable not only as an optical instrument but also as an information processor.

Embryological studies show that the eyes begin to develop very early as two small buds on the neural tube that eventually becomes the brain. These buds grow outward as stalks during development, form the optic nerve, and finally reach the inside of the skin, the cornea, and become the retinas. Thus there is every reason to regard the retinas as the forward outposts of the vertebrate brain.

Color

3

Color is an immediate, automatic part of our perception of the world, and it pervades our experience. It is often the first characteristic we choose to distinguish one item from another. "I'll wear the brown suit tonight." "It's the house with red shutters." Subtle variations in color supply us with vital information. A pallor shows in the infant's cheek and the mother knows he is ill. The vigor of a leaf, the fertility of the soil, the taste of a peach—all are judged by color. It is our metaphor for sound, for music, for personality. The color of eye, of hair, or of skin can lead to love, fear, hate, and war.

The ancients debated whether color was an attribute of light or of the object. We now resolve this debate, as we so often do with scientific debates, by saying the answer is "both." White light is made of many colors—indeed, we can produce colored light from it—and an object reflects some of these colors and absorbs the rest. Yet color is more than the light and the object, as Newton himself recognized, for ultimately it is a phenomenon of perception. It is a creation of the eye and the mind. Thus the scientific study of color, which began in the nineteenth century and continues actively today, lies not in physics or in physiology but in the area overlapped by these sciences and the psychology of perception.

The Parameters of Color

One would be tempted to say that the discovery of the wave nature of light provides the explanation of color. For each wavelength within the range to which the eye responds corresponds to a different color; and we see all colors in the continuum from violet and blue at the shorter wavelengths to red at the longer. We do not have words in our language for all the colors we perceive. Indeed many languages have fewer or more words than we do. But it is reasonable to assume that all humans are born with approximately the same physiological equipment with respect to sight and that we perceive the colors of the rainbow in the same way. Yet it is clear that the rainbow, or the spectrum of a prism, does not exhaust the possibilities of color. For there is no brown, no purple,

49

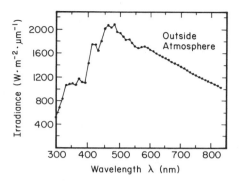

Fig. 3.1. The spectral curve of the sun. The intensities of solar energy in a small wavelength interval around wavelength, λ, are plotted as a function of λ. (Adapted from Günter Wyszecki and W. S. Stiles, *Color Science, Concepts and Methods, Quantitative Data and Formulae*, [New York: John Wiley and Sons 1982], 6.)

no black, white, or gray, to say nothing of pink, mauve, apricot, or any of the hundreds of other colors created by manufacturers of textiles, paints, and plastics, as well as by nature. (One may argue about whether white, black, and gray are colors. But seeing a gray or a black object is surely a perceptual experience of the same order as seeing a yellow object; the argument, finally, would be just semantic.)

The question we should ask first is not what can be classified as a color, for this is only a matter of words, or how many colors there are, but what are the parameters with which color may be defined. Consider white light from the sun, for example. It is not enough to say that this light contains colors of all wavelengths from violet to red. To specify the nature of sunlight we would have to say how much light there is of each wavelength. It turns out that the greatest intensities are in the green and blue portion of the spectrum; the shorter- and longer-wavelength regions are less intense.

Intensity, to be specific, is defined as the amount of energy arriving each second at a receiver of a fixed cross-sectional area (say, 1 square meter). The receiver itself has a response—the darkening of photographic emulsion or the rise in temperature of an absorptive surface—which may also vary with wavelength. More modern techniques involve the photoelectric effect, that is, determining the electric current generated in a photocell by light of different wavelengths. Knowing the response properties of the detector, one can unravel the intensity-wavelength variation of the light source, its *spectral curve*. It is important, however, to bear in mind that intensity is a property of light alone and is distinct from the response of the light detector.

Figure 3.1 shows the spectral curve of the sun, recorded above the atmosphere by NASA using equipment in rockets and high-flying airplanes. Light from an incandescent bulb also contains contributions from all wavelengths, but the intensity of the yellow-to-red region is greater than that at shorter wavelengths. Light passed through a green filter may contain predominantly green, smaller amounts of yellow and blue, and little orange or red. In other words, to specify completely the color of a beam of light we must present a series of numbers giving the intensity for each wavelength in the visible range. In principle the parameters necessary to define a color seem to be infinite in number. (In practice a large but finite sampling over the visible would suffice.) We may say that the "dimensionality" of a color seems to be infinite. A spectral color, one selected from a narrow band of the prism's spectrum (red of 650 nm, for instance), is a special case, in which all the intensities are zero except for the intensity at this one wavelength.

Yet we must be careful to distinguish the construction or measurement of a light source in the laboratory from the perception of color by a human observer. For although the physicist may analyze light by determining the relative strength of each wavelength, it is not necessarily true that the visual system perceives light in this way. Indeed systematic color perception experiments, begun in the nineteenth century (James Clerk Maxwell was among the pioneers in this research) and continuing to the present time, have revealed far more subtle and interesting mechanisms for the processing of color by the eye.

Suppose we project spectral yellow light on a screen, which then reflects yellow to the eye. Now let blue light shine on the same spot on the screen (fig. 3.2). The combination of blue and yellow is white.[1] Similarly, a red light added to a blue-green spectral light (say, around 490 nm) is seen as white. Red combined with a spectral green at around 510 nm is seen as yellow. There does *not* exist a one-to-one correspondence between the set of numbers giving the intensities of various wavelengths in mixed light and the perceived color of the light. Rather we find that some entirely different mixtures of lights can produce precisely the same subjective response.

Now suppose we mix blue and red light in varying proportions. We produce a continuous range of purples, from reddish blue to bluish red, each color retaining some psychological aspect of both red and blue. On the other hand, mixing red and green in fairly equal proportions produces, as mentioned, yellow, a color that we do not in any way see as reddish green and that does not seem to possess the attributes of its component colors. No combination of red and green will produce white but adding a fraction of blue to a red-green mixture will do so.

The results of the color-mixing experiments may be summarized by the following far-reaching statement, embodying what

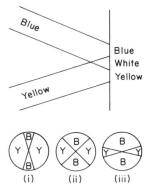

Fig. 3.2. Color-mixing experiments. (*top*) Lights can be mixed by shining colored light on a white screen. Where yellow and blue beams overlap on the screen the illumination is white. (*bottom*) Three color wheels are painted with sectors of yellow and blue. When the wheels are rotated rapidly the colors are added in the eye; you see each wheel as a uniform color: (*i*) an unsaturated yellow; (*ii*) white; (*iii*) an unsaturated blue.

is called the *trichromatic theory of color:* Within certain limitations, to be discussed below, any colored light source (a "test" source) whatsoever may be *matched* by a suitable combination of the three colored lights—which we will tentatively call *primary*—red, green, and blue. The significance of matching in this context is that these experiments, while relying on subjective responses, nevertheless do not require subjective interpretations of the words for colors. A subject is given a test light in his visual field and three knobs controlling red, green, and blue lights that are superimposed in a region adjacent to the test source. He varies the intensities of the three primaries until the light they produce cannot be distinguished from the test source. The results are not only that all observers succeed in this task but that the intensities which they arrive at are the same for different observers, and for the same observer at different times. In other words, we have uncovered an inherent attribute of the human color perception system, an attribute which in some sense has three parameters. Three colored lights are necessary and sufficient to match the test source; two lights—at least for humans with normal vision—will not do. Another way to express this is to say that color, as a perceptual phenomenon, is three-dimensional.

Two colors that are spectrally different (that is, made of different combinations and intensities of wavelengths) but appear the same to the eye are the same in other respects too. If a certain yellow combines with blue to produce white, and if we can match that yellow with a combination of red and green in a given proportion, then that combination of red and green will act with blue in the same way to produce white.

The trichromatic theory of color, that any color can be matched by a combination of red, blue, and green, is both weaker and stronger than the statement of it made above. Certain colors—for example, a spectral yellowish green at around 540 nm—cannot be matched by a spectral red, green, and blue. But we can restore the principle of the trichromatic theory if we extend the definition of matching in the following way: Picture the yellowish green light on one spot of a screen, and on an adjacent spot to the right place a combination of red and green. Adding blue to the spot on the right does not permit us to match the two spots, no matter how we vary the intensitites of the three primaries. But if we add blue to the spot on the *left,* then we can, by adjusting the intensities of the primaries, match the two spots (although neither will be the same as the original yellow-green). In this more general sense, red, green, and blue are primary; starting with any color on the left spot we may produce two matching spots by using appropriate combinations of the three—red, green, blue. The generalization is sometimes referred to as "negative" mixing, in that

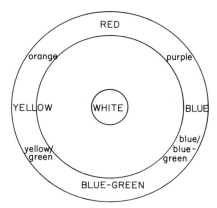

Fig. 3.3. The color circle. White is at the center because it is a combination of all wavelengths. Colors around the rim are spectral (except for purple); those opposite each other are complementary. Between the center and the rim are the desaturated colors for each hue. (Often versions of the color circle are drawn with red, yellow, green, and blue at four equally spaced points on the rim.)

adding blue to the spot on the left may be thought of as equivalent to adding *negative* blue on the right.

Having extended the definition of mixing in this way, we can state the trichromatic principle in a much stronger way. The primary colors do not have to be taken as red, green, and blue. Rather, *any* three colors can be taken as primaries, and by a suitable combination of these (allowing for negative mixing) we can match any color at all. Thus, the question of which colors are the primary ones becomes empty. Any three colors may be taken as primary,[2] although we will see that practical considerations render some choices more convenient than others.

The definition of white is somewhat arbitrary. We can take it to be the color of sunlight or, more commonly, the color of skylight on a cloudy day or the color of an artificial light that approximates sunlight. A special case of color mixing is the determination of pairs of colors that may be added to produce white. Such colors are called *complementary*. We have already seen several examples, and for most spectral colors there is another spectral color that is complementary to it. There is, however, a range of wavelengths, in the blue-green to yellow-green region, for which the complementary colors are various shades of purple. These purples are not spectral colors, because they are made by mixing red and blue, which are at opposite ends of the spectrum. No single wavelength gives these sensations of purple. (In contrast, the sensation of orange can be made by mixing red and yellow lights, say red at 620 nm and yellow at 575 nm; but orange is also seen in pure spectral light at around 590 nm.) Nevertheless, because purple acts as a complementary color in this way, one can represent the colors as a closed continuum, the color circle, red running through yellow, green, blue, and then through purple back to red again. White is placed at the center, midway between the various pairs of complementaries (fig. 3.3).

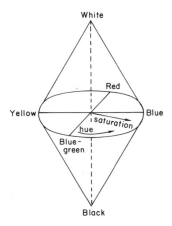

Fig. 3.4. The color solid. The color circle may be extended to three dimensions by imagining a color solid, with brightness measured along a vertical axis. The central axis includes the achromatic colors, from white through the grays to black. Hue is the variable measured along the circumference of any circle about the axis; complementary colors again are on opposite sides of the achromatic axis. Saturation is measured as distance from the axis toward the outer circle. In some representations the circle is tilted with the yellow edge up, the blue edge down, because yellow is perceived as brighter, and blue darker.

Hue, Saturation, and Brightness

The location of white at the center of the color circle suggests another way to classify colors. For if a given amount of blue and a given amount of yellow produce white, then we can increase the blue intensity slightly, decrease the yellow, and thereby produce a color we might call whitish blue. That is, it is located in the interior of the color circle, not at the center, but displaced slightly toward the blue region of the perimeter. If we increase the blue still further and decrease the yellow further, we find again a whitish blue, but now still closer to spectral blue, closer to the perimeter of the color circle. Thus there is a dimension of color associated with the degree of purity of the light, the extent to which it is near the perimeter of the color circle and so spectrally pure, rather than mixed with many other wavelengths. This dimension is called *saturation*. A light source in which blue is the predominant wavelength, containing a substantial admixture of other wavelengths (the sky is a good example), is desaturated; pure spectral blue is highly saturated.

For most light sources, although they may be mixtures of many or all wavelengths, observers will consistently perceive a dominant color. We call this dominant color the *hue,* and it is hue to which we give names like red, blue, or greenish yellow. Having established the hue, we may ask to what extent that hue appears to be mixed with white or, perhaps, how much hue is there in the total visual sensation. (Your eyes are bluer than lapis lazuli!) Under experimentally controlled conditions people give fairly consistent responses to questions like these, and thus there can be established the perceptual dimension of saturation.

Finally, since color means the total perceptual response to light, there is certainly another dimension needed to describe it— the *brightness,* which is related to the total intensity of the light wave. These terms may be made more precise, as we will see shortly, but the point here is that we again find color to be three-dimensional. In terms of color-mixing experiments we saw that a color would be specified by giving the intensities of three primaries needed to produce the color. Alternatively we may give hue, saturation, and brightness (fig. 3.4). Color is a three-dimensional quantity and we have found two entirely different ways of expressing those three dimensions.[3]

Whites, grays, and black are part of our system of color; they can be thought of as completely desaturated versions of any color. And the distinction between one gray and another, on a continuum from white to black, is given by brightness alone.

The three color variables, while closely related to the physical properties of light, must be thought of as *psychophysical* variables. They involve not only measurements of intensity and wavelength, but also the judgment, the subjective response, of human observers. Subjects, for example, may view two spots of red and blue spectral light, wavelengths fixed, and be asked to vary the intensity of the blue (by turning a dial) until the two spots are equally bright. Or subjects may vary wavelength with fixed intensity or mix spectral lights with each other or with white light and be asked to match spots according to hue or saturation. Hue is primarily correlated with the dominant wavelength, and brightness with intensity; but it is found that to a small extent we perceive hue to depend on intensity, and brightness depends significantly on wavelength. If subjects are matching hues, they will match an intense light in the red region with a less intense light in the orange. Given blue and yellow of equal intensity, the yellow will be called brighter. Similarly, saturation depends to some extent on the physical properties most closely associated with hue and brightness. In the final analysis, as we have said, color is a property of perception; and the consistency of results in psychophysical experiments, using many subjects and the same subjects over a span of time, testifies to the psychological reality of the color variables. To understand the how and the why of trichromatism, we must follow the message of light into the eye and as far as possible back to the brain.

Color Receptors in the Eye

One of the early important observations about color perception was made in 1825, by the Bohemian Jan Evangelista Purkinje. Watching the flowers in his garden at twilight, he saw that the relative brightness of differently colored surfaces shifted as daylight diminished. As the sky darkened, red flowers quickly became black, while in twilight the green leaves were still bright. What he experienced was the gradual change from the mechanism of daytime vision, the cones, to the mechanism of night vision, the rods, and in particular the fact that the sensitivity of the rods, besides being generally greater than that of the cones, also varies with wavelength in a different way (fig. 3.5). The relative sensitivity of rods at long wavelengths (compared with the sensitivity at short wavelengths) is less than that of cones, and so, relative to other lights, we lose the perception of reds more quickly as it gets dark.

A physiological basis for the trichromatic nature of color vision was suggested by Thomas Young in 1801, even before the color-matching data existed. He proposed that the eye possesses three types of receptors, each with different wavelength sensitivity, that is, three types of cones for three-dimensional color vision. The argument leading to this conclusion is rather more subtle than it seems on the surface, and it deserves closer analysis.

Suppose, for example, there are three receptors, sensitive to red, green, and blue, respectively. It is clear that sensitivity cannot be limited to three small wavelength regions, say, red at 650 nm, green at 510 nm, and blue at 475 nm, for then a spectral orange at 590 nm would simply not affect the eye at all and would appear black. Certainly, each receptor must have a range of sensitivity such that at least one receptor responds for each wavelength in the visible spectrum. Now suppose there is a red receptor that is sensitive, to varying degrees, in the region from 575 to 675 nm, while the blue and green receptors are insensitive in this region. Let us say it has maximum sensitivity at 625 nm and half maximum sensitivity at 575 and 675 nm. Such a system would respond to both red light at 625 nm and orange at 575 nm. But how would the eye distinguish red from orange? The orange signal, one might think, would be half as strong as the red signal. But how, then, would one distinguish an orange light from a red light of half the intensity?

The point here is that the cells of the visual system, in common with most nerve cells, have only one way of responding. The photoreceptor can change its voltage, the ganglion cell can change its firing rate, but each cell can convey no other information about

Fig. 3.5. The luminous efficiency curves for rods (*upper curve*) and cones (*lower curve*). The rods' sensitivity, arbitrarily set to 100% at maximum, is much greater than the cones'. The rod peak is at a shorter wavelength. These curves show the response of a "standard observer" as defined by the International Commission on Illumination. They represent response to light of various wavelengths under conditions in which either rod vision or cone vision predominates. (They are not direct measurements of light absorption by the pigments.)

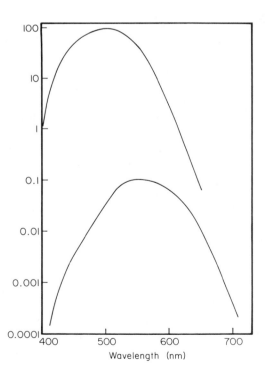

what caused its response. The cell's response, in other words, is one-dimensional. There is no way for it to distinguish a strong signal in an insensitive part of the spectrum from a weak signal in a sensitive part of the spectrum. (This is in fact the situation in night vision, where, using only rods, we see various shades of gray. Often we perceive the same gray coming from objects which, in daylight, are different in both hue and brightness.)

What nature must do, therefore, is provide overlaps among the three types of color receptors in their *spectral sensitivity functions* (the dependence of sensitivity on wavelength). Moreover, in a given region of the spectrum there must be at least two types of cones operating, in such a way that the spectral sensitivity functions of the two are different over that region. Under these conditions, when we move from, say, 625 nm to 575 nm, the *relative* responses of the two receptors will change; and there is no way in which that change can be mimicked by a change in intensity. Thus (although not all details of the mechanism are fully understood) the neural system can distinguish red from orange.

So much for theoretical deduction. In the twentieth century new types of experiments, using sophisticated electronic and physiological techniques, have succeeded spectacularly not only in verifying the three-cone theory of color vision but in mapping the sensitivity function of each type of cone. For one thing, one can obtain surgically the retina of an experimental animal. Goldfish have three-color vision and have been extensively studied, as has the macaque monkey, which has color vision very similar to that of humans. Moreover, there are some pathologies in the human which require removal of an eye, and from this source measurements have been made on human visual cells. Through the optical system of a microscope, used in reverse, light is focused onto a very small area; a monochromatic beam can be directed through the outer segment of a single cone. By studying the transmission of light through the cell as a function of wavelength, one determines the absorptive capacity of the pigment. These *spectrophotometry* experiments, carried out by E. F. MacNichol, Jr., George Wald, Paul Liebman, and others in the 1960s, represent a considerable technical feat. The cone is small (about 0.0001 cm in diameter) and contains an extremely small quantity of visual pigment.[4] The intensity of the beam must be low so that the cone pigment is not bleached, and thus very sensitive photoelectric detectors are needed. The fact that cones, like rods, can be bleached is an important element in these experiments, for the absorption of light by the retina is not due solely to the visual pigments; other cells and other parts of the cone absorb light. Therefore, one first makes an absorption measurement, then bleaches the cell so that the pigment is transparent, and while the cell is still bleached one makes another absorption measurement. The ratio of these two measurements represents absorption by the pigment alone.

A second type of experiment, *retinal densitometry,* pioneered by W. A. H. Rushton, allows measurement of the properties of the visual pigments in a living human. These experiments rely on the fact that when light has passed through the retina it reaches the pigment epithelium, which, although highly absorptive, does reflect a small fraction of incident light. This fraction passes through the retina again and exits through the pupil. Thus, although one cannot focus on an individual cell, one can measure the absorption of light on passing twice through the living human retina. In these experiments one makes use of selective bleaching, in which a bright light may render two types of cones temporarily inoperative, so that one studies only the third type of cone in the subsequent minutes.

The conclusions of these experiments are that there are, indeed, three types of cones, each containing a different type of pigment. The absorption curves of these pigments are known (fig. 3.6). Each has a maximum in a different part of the visible spectrum, and they are usually labeled blue, green, and red, in order of increasing wavelength of the peak. The curves are broad and indeed show substantial overlap. Although the "red" cone is certainly the most sensitive of the three in the red region, its maximum sensitivity is in the yellow.

Some of Rushton's experiments were with color-blind subjects, and his work has succeeded in illuminating something of the nature of this abnormality. Over 8 percent of the population, mostly males, is color-blind, usually lacking to some degree the ability to distinguish colors along the red-green axis. Most color-blind individuals possess three different cone pigments, but the spectral absorption curves of their pigments are in some way anomalous. They therefore respond differently to various hues and perform differently on color-matching tests. A minority simply lack either the red or the green pigment entirely. Possessing two pigments, their vision is *dichromatic* rather than trichromatic. On color-matching tests they can match any color by a suitable combination of only *two* "primary" colors, although colors that they perceive as matched would not necessarily be matched by a subject with trichromatic vision. Since the two pigments possessed by dichromats are normal, scientists have been able to use color-blind subjects to aid in the study of normal color vision.

What the dichromat has done, in effect, is to discard, or lose, one of the three pieces of information that the trichromat makes use of in interpreting light. But it is worth recognizing that the trichromat has also discarded a great deal of information, since he reduces the infinite-dimensional character of the physical

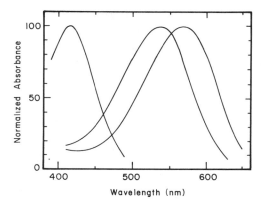

Fig. 3.6. Absorption of light by the three different cones as a function of wavelength. For the sake of comparison, the curves are adjusted so that their maxima are at the same height. Substantial overlap of these curves is apparent; for example, light at 480 nm is absorbed by all three pigments. Note that the amount of light absorbed is not necessarily a measure of the extent to which that light affects perception. (Redrawn from J. D. Mollon and L. T. Sharpe, eds., *Colour Vision, Physiology and Psychophysics* [New York: Academic Press, 1983], 60.)

light spectrum to the three-dimensional retinal response. In this sense we are all color-blind to a very large degree. In fact what the color-sensitive person does is to formulate, somewhere along the way from the retina to the brain, certain *abstractions* constructed from the information in the light, abstractions to which are given names like "hue," "saturation," "yellow," and "blue."

Why is nature so wasteful? Notice that in the case of sound we accept much more spectral information. Sound waves of many different wavelengths will be heard as different pitches (except for some special cases) and will not be processed into a single sound sensation. In a rough sense, the ear has different receptors for many different wavelengths. But we have only two ears, and consequently our determination of the direction of a sound source is not extremely precise. In vision, on the other hand, our measurement of direction is highly developed and very important. We need the big two-dimensional map of the world in front of us, and so we have different receptors in the retina for many different directions in space. With millions of receptors forming the map, perhaps nature could not provide each with a hundred wavelength channels.

Furthermore, nature is usually economical. Abstractions, constructed by a finite brain out of the raw data of perception, may be the most efficient way for that brain to avoid being overloaded and to direct our response to the environment.

Determination of the sensitivity of the three cone systems provides a physiologically real three-dimensional representation of color. Suppose we are given an arbitrary light with some complicated distribution of wavelengths—neither a spectral color nor white. We may measure the total response this light will produce in the red cone pigment, taking account of the fact that that pigment responds to a broad range of wavelengths. This procedure gives us a number, characteristic of that light. Similarly we obtain two other numbers pertaining to the green and blue pigments. The set of three numbers will uniquely describe not the light but the color. A quantitative representation could equally well be created based on hue, saturation, and brightness, although it would not be derived from physiological measurement. (In principle, cells responsive to these properties may exist in the brain and may be discovered and mapped at some time in the future.) Some system of this sort, specifying color quantitatively, would clearly be useful, particularly to industries that must produce, sell, and buy materials of given colors.

The Chromaticity Diagram

In 1931 the International Commission on Illumination met to establish such a system. At that time the spectral properties of the human cones were not known, although the three cone pigment theory was widely known and believed. Furthermore, the color-matching data existed, and this placed certain limits on the response curves. For if two different wavelength distributions are matched—seen as the same color—then the quantitative color system must be arranged so as to assign the same set of three numbers to the two distributions.

    The commission invented (or defined) response curves for three hypothetical cones, consistent with the data, and in terms of these formulated the CIE (for the French title, Commission International de l'Eclairage) three-dimensional coordinate system of color. Although the three coordinates are not directly connected with the responses of real human cones, they are nonetheless useful as a reproducible description of color: each set of three numbers corresponds to a given color. The chromaticity coordinates describe light, but they are often used to characterize a colored surface (say, of manufactured goods). In this context they refer to the light reflected by the surface when it is illuminated by

Fig. 3.7. The chromaticity diagram. Each point within the chromaticity diagram represents a different color, labeled by the two coordinates x and y. The curve (the spectral locus) represents the spectral colors, from red through orange, yellow, green, and blue to violet. The curve is deduced from color-matching experiments; the color labels are added on the basis of hue-naming experiments, in which subjects are asked to give names to different stimuli. For example, the purest red is located not quite at the long-wavelength end of the spectral locus but is displaced slightly down the purple line. White is in the central region of the diagram; the white point is not uniquely defined.

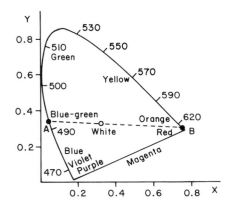

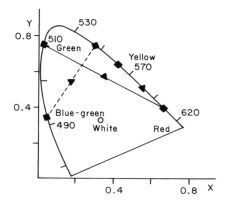

Fig. 3.8. Mixing of spectral colors. The long-wavelength end of the spectral locus, above about 540 nm, is fairly straight. Hence mixing red and yellow produces a saturated orange. On the other hand, mixing red with green (at 510 nm) produces a desaturated yellow because the spectral curve bends sharply toward 510 nm. Mixing yellow-green with blue-green produces a very desaturated green.

white light. Since there is a range of lights that would be called "white," the CIE has arbitrarily chosen a standard white light. Of course if the illuminating light is different, then the reflected light will also be different.[5]

For many purposes the absolute brightness of a color is of less interest than hue and saturation, and so one of the three coordinates can be eliminated. The two remaining coordinates, x and y, may be conveniently plotted in a two-dimensional graph known as the *chromaticity diagram* (fig. 3.7), each point in the x-y plane representing a unique color.

To illustrate the utility of the chromaticity diagram, we first consider all of the pure spectral colors, from red to violet. Each of these is represented by a point in the x-y plane, and the full spectrum forms a curve, the *spectral locus*. When two colors, represented by points A and B in the diagram, are mixed, the resulting color lies along the straight line between A and B; its proximity to A or B is a measure of the relative intensities of colors A and B in the mixture. Thus white is seen as midway along the line between red at around 650 nm and blue-green at 492 nm. If red is stronger than blue-green in the mixture, the resulting color is at a point between white and red, and it is seen as an unsaturated red (fig. 3.8). Any visible light must be some combination of the spectral colors, and so the points inside the spectral locus represent all the lights, and the only lights, that can be produced. The straight line drawn between the red and violet ends of the locus represents the purples, which, although not in the spectrum, nevertheless complete the outer boundary of the colors that can be seen. This entire outer boundary corresponds to the maximally saturated colors; points in the interior, closer to white, are less saturated.

The chromaticity diagram helps to explain the significance of primary colors. Any three colors (spectral or not) form a triangle in the color space, and the colors that can be produced by

Fig. 3.9. Primary colors. Three primary colors are represented by three points in the chromaticity diagram. The points form a triangle. All colors inside that triangle can be made by positive mixing from these primaries. The choice of red, green, and blue fails to give a saturated yellow. The choice of red, yellow-green, and blue gives a better yellow, but a less saturated green. All hues are possible with either choice of primaries. But no set of primaries can give all *colors*, since any set will miss some of the highly saturated colors near the spectral locus.

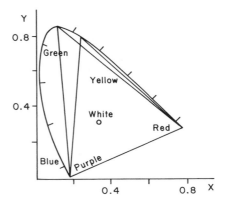

mixing these primaries are those in the interior (and along the edges) of that triangle (fig. 3.9). It is clear that no triangle will include all points within the spectral locus, and so no set of three primaries can match every possible light. Only with negative mixing in the sense described earlier (which takes us outside the triangle) can all colors be matched. If we are restricted to positive mixing of three lights, the best we can do is to choose three primaries that will give us the widest possible range of colors among the mixtures. A glance at figure 3.9 shows that this choice is spectral red, blue (at around 470 nm or less), and green (at around 520 nm). This has the disadvantage, however, that although all hues could be mixed, the blue-green would be fairly unsaturated. Moving the green primary down to 500 nm improves the blue-green mixture but severely desaturates the yellow-green.

The Mixing of Lights and Pigments

Choice of primaries becomes a practical matter when we seek to design versatile sources of colored light. Color television pictures are produced by placing behind the television screen three kinds of phosphorescent materials, each glowing a different color when struck by the cathode-ray beam. The broadcast signal determines which phosphors will glow at different points along the screen. The eye adds up the many little points of color and perceives all colors generated by the set. Some compromise must be made because phosphors at the longest wavelength of red and at the shortest wavelength of blue are not very intense. So to avoid a loss in brightness, the primaries for television are a somewhat orange red, a slightly unsaturated blue, and yellowish green. Some loss in fidelity, particularly to the spectral colors, is inevitable.

A similar sort of color addition has been used by artists in mosaics, and in painting since the time of the French impressionists. An artist may paint many small strokes of a variety of colors such that, at a distance, the eye sums these into a smoothly varying pattern of color. Viewed up close one sees, for example, dabs of purple, yellow, or green in a painted face, yet from a distance the subtle tones of the skin are rendered faithfully. The landscapes of Claude Monet show these methods dramatically, and color mixing is brought to an almost scientific precision in the pointillism of Georges Seurat. Seurat's canvas is covered by thousands of dots of uniform size, which are blended by the eye into smooth gradations of color. Pointillism as a theory and technique was not long lasting, but it influenced Henri Matisse, and it can be felt in the choppy strokes of color in the paintings of Van Gogh.

On the other hand, most color mixing in painting—and in industry as well—is based on what is called a *subtractive* process rather than on the adding of lights. The artist normally mixes pigments to create the shade that he needs. A pigment does not produce light; rather it absorbs light. A red pigment absorbs all the colors in white light except for the red. And so the laws of mixing pigments are entirely different from the laws of mixing lights. As an illustration, if green and red pigments are mixed, the green pigment absorbs all light except green, the red all except red, and so the mixed pigment absorbs all light in the spectrum. It appears black, or at least very dark; certainly not yellow. On the other hand, the reflection of light by pigments is really not highly selective. A blue-green pigment absorbs yellow, orange, and red and reflects blue, green, and violet. Yellow pigment is yellow in hue but reflects substantially over the range green-yellow-orange-red; it absorbs blue and violet. When the two pigments are mixed, the result is green, the only color *not* absorbed. But mixing the blue-green with red leaves very little light to be reflected. If the blue-green pigment is diluted in the mixture, yellow, orange, and red will not be absorbed completely; the weak reflection of these colors, under illumination by white light, makes the mixture look brown.

Here is a new aspect of color mixing that is important in both art and industry: if we think only of hue and saturation, we forget the fact that a color can be so dim that we see it only as black. Indeed the appearance of a colored area depends not only on the light we receive from that area but also on the surrounding area. In most color-mixing experiments a colored light is presented to the eye against a black background. But in the everyday world we view a colored surface against a background of walls, floor, grass, and sky, and these affect our perception of the color. We might think that the difference between a gray and a white surface is simply that the gray surface sends less light into the eye than the white. Yet, if we look at a white sheet of paper in a dimly lit room, it will send less light to us than will a gray sheet of paper held in the sun. Nevertheless we unambiguously call the gray paper gray and white paper white. We can make this interpretation because we are aware of light from the surrounding areas.

A circle of color painted on a white background and viewed in white light is characterized not only by its hue and saturation but also by its brightness. This in turn is related to the percentage of the incident white light that it reflects. If this percentage, the *reflectivity,* is low, we get what the artist calls "dull" colors. If hues in the orange and yellow region have low reflectivity (less than about 20 percent), they appear as brown. Reds of low reflectivity are maroon; dull greenish yellow appears as olive green. Is there such a thing as "brown light"? Brown is a sensation, and it can be produced by light in the red-to-yellow region, but only if that light is surrounded by a brighter contrasting light.

The artist creates dull colors by adding black pigment, reducing the reflectivity of his mixed colors. More difficult is avoiding dullness. For lightening a pigment by mixing it with white sacrifices saturation. Mixing pigments can produce many interesting hues, but, as generations of children have discovered, the most promising experiments tend to come out muddy brown. To produce a pure green the painter wants a pigment that reflects only a narrow wavelength band around, say, 510 nm. In fact, as we have seen, pigments reflect over broad bands. But even if it were possible to select a narrow band, the result would be poor, because only a small portion of the incident white light would be reflected; the color would be very dark. In practice there are many bright pigments, but because they reflect over a wide wavelength region,

Table 3.1   Subtractive Mixing Using Colors Complementary to the Primary Lights (Red, Green, and Blue)

Pigment Color	Its Complement	Light Reflected or Transmitted
blue-green	red	blue and green
magenta	green	red and blue
yellow	blue	red and green
blue-green plus magenta	red and green	blue
blue-green plus yellow	red and blue	green
magenta plus yellow	green and blue	red

Note: Light transmitted by a blue-green dye is white minus red, therefore light equivalent to a mixture of blue and green (similarly for magenta and yellow). Therefore, light transmitted by a mixture of blue-green and magenta contains only blue. In this way the three pairs of dyes provide blue, green, and red lights.

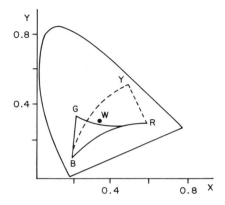

Fig. 3.10. Mixtures of four tempera in the chromaticity diagram. Point *G* is an emerald green pigment; *B* is ultramarine blue; *Y* is yellow; *R* is red; *W* is white. The curve between two of these primary pigments indicates the result of mixing the two. Two choices for a set of three primaries are shown. Since the curve between red and green falls below the white point, there is no way in which the hue yellow can be produced with the primary set red-green-blue. (R. D. Edge and R. Howard, *American Journal of Physics* 47 [1979]: 145; copyright AAPT.)

their colors may be unsaturated. In the case of yellow, a glance at the chromaticity diagram shows that the spectral locus is very close to a straight line. This means that reflected light over a wide wavelength band, say from 540 to 620 nm, will add up to a color very close to spectral yellow, perhaps at 580 nm. Thus a yellow pigment can be bright and as saturated as a spectrally pure yellow.[6]

Artists select a broad-band yellow as one of the three "primary" pigments or dyes. The other two are a magenta, midway between red and violet, and blue-green. These pigments are the complements of the three primary lights (Yellow is complementary to blue, magenta to green, and blue-green to red.) When illuminated by white light, each pigment absorbs one of the primary lights and reflects the other two. As seen in table 3.1, a mixture of *two* of the pigments reflects *one* of the primary lights. Hence, by the proper mixture of the three pigments, we can produce reflected light that contains the desired combination of red, green, and blue. Such processes are used in inks (four-color reproduction is based on black, yellow, magenta, and blue-green) and in photographic film (where three separate layers of emulsion provide three-color reproduction).

Bear in mind, however, that this is an oversimplified description of mixing. In fact, each pigment is characterized by a detailed absorption curve—the degree of absorption for each wavelength—and only if this curve is known for each of two pigments can the results of subtractive mixing be predicted. Furthermore, mixing of coloring agents is not necessarily subtractive. It can be additive (as with mixed lights) if pigment particles act like a mosaic across the colored surface; or it can be a more complex process, neither additive nor subtractive. More frequently, the result of mixing is determined by experiment, or by trial and error. We are far from being able to create all colors by mixing three primaries, although we *can* expect to produce all hues.

Figure 3.10 summarizes an illustrative experiment by R. D.

Edge and R. Howard, in which they investigate primary pigments in opaque poster colors (tempera). One set of primaries was red-green-blue and another set was red-yellow-blue. The mixed colors were determined by measuring the reflection of white light and were plotted as points on the chromaticity diagram. The curve between blue and yellow shows the results of mixing these two pigments in proportions from 100 percent blue to 100 percent yellow; similarly for the other two curves. If mixture were additive, the curves would be straight lines. The white point is shown in the middle. We see that the red-yellow-blue combination produces all hues, but most colors are fairly unsaturated. The red-green-blue primaries do not produce yellow at all, since the curve from red to green lies on the wrong side of white.

The Opponent-Color Theory

Much more can be written about the interplay of hue, saturation, and brightness in the mixing of pigments. But instead let us return to the trichromatic theory of light and the three primary *lights,* red, green, and blue. For there is something unsatisfactory—and you may have felt it—in the suggestion that this trio encompasses all color vision. Red-green-blue does not seem to provide a complete sampling of the range of color vision (as bass-tenor-alto-soprano covers the range of the human voice). It is true that all hues are generated by appropriate mixtures of red, green, and blue light, but there is a lack of symmetry in our psychological response to these mixtures. If we mix blue and green in varying proportions, we see a continuous range of blue-green colors, all possessing some of the feeling of blue and green; and similarly, mixing red and blue, we see the purples, all possessing the feeling of red and blue. But when we mix red and green, we have, not reddish-green, but yellow. We feel yellow as an entirely different sensation, psychologically independent of red, green, and blue (unlike orange, say, which clearly retains a feeling of yellow and red).

Such feelings are subjective and difficult to categorize scientifically. Yet they are real and led the nineteenth-century physiologist Ewald Hering to propose an alternate theory of color based on four fundamental colors: red, yellow, green, and blue. This is known as the *opponent-color theory,* because the four colors are associated as pairs, red-green and yellow-blue, with the members of a pair thought of as acting in some sense in opposition to each other. (A third pair, white-black, was included by Hering.) For one thing, yellow and blue are complementary colors, adding up to white; red and green (if we take a bluish green) are also complementary.

Furthermore, these pairs of colors act in an opposing way in various kinds of color-contrast observations. The most familiar of these is the phenomenon of *afterimages*. Everyone has experienced dark and light afterimages. If you stare at an object with sharply contrasting black and white areas and then quickly turn away and look at a white wall, you see a negative afterimage, a similar pattern in which white is black and black is white. This can be explained as fatigue of the retina in those areas where the original object was bright. (The exact mechanism of this fatigue is not fully understood, but it is known that it is not simply due to bleaching of the photopigment. Rather, an induced temporary insensitivity in the neural connections beyond the photoreceptors must also be involved.)

Color afterimages occur in a similar way. If you stare for 15 seconds or so at a bright color and then turn to a white wall, you see an afterimage in the complementary color. You can produce an afterimage with a glance at an incandescent bulb. Do not be surprised to see several small circles of blue; you see several images because the eye constantly makes rapid jerky movements. (In normal vision the brain compensates for this motion and we are not aware of it.)

Afterimages are usually observed by looking at a white wall after adapting the eye. But the tendency to see the complementary color is quite general. For example, if the eye is adapted to a red light and then we view light of any particular color, the perception of that color will be shifted in the direction of blue-green. If the second light is blue-green, it will appear more saturated; if the second light is yellow-green, it will appear a desaturated green.

These effects can be easily explained on the basis of the three-cone system of vision, in the same way that dark-light afterimages are explained. The adapting light fatigues that portion of the retina sensitive to it, so that when we view a second light the nonfatigued portion dominates the view. Not so easily explained by retinal response are simultaneous-contrast experiments, in which a given patch of color is viewed against a surrounding field of another color. We have already commented on how the surrounding field affects light and dark. There are similar effects on hue and saturation. In general, the color of the patch of light is shifted toward the color complementary to the surrounding light. A circle of green light surrounded by red will be seen as bluish-green, shifted toward the complement of red; the same patch of light appears green against a dark background. Subjects in these experiments, as in color-mixing, are not asked for verbal descriptions of colors but rather must match colors with standards in the same field of view.

Simultaneous-contrast effects (as opposed to afterimage effects, which may be thought of as successive contrast) cannot be explained by fatigue of a specific portion of the retina and may well have to do with horizontal connections in or beyond the retina.[7] These considerations, along with the dark-light contrast effects, led to Hering's color theory. The theory did not receive a great deal of attention in the nineteenth century, to some extent because of the great prestige of the scientists (such as Maxwell and Helmholtz) associated with the rival three-color theory, but also because it had no systematic experimental foundation. Within the past thirty years, however, exciting new experiments, intimately probing the visual pathways, have provided striking evidence for the opponent-color theory. It is not that the three-color theory has been superseded; for there is no doubt that located in the retina are the three types of visual receptors that we have

Fig. 3.11. Oscilloscope records of responses of spectrally opponent cells in the LGN of the macaque monkey. At left lights are off. When a light is turned on there is inhibition if the wavelength is 480 nm or less and excitation in the wavelength region from 505 to 633 nm. (Reprinted by permission, from R. L. De Valois, I. Abramov, and G. H. Jacobs, *Journal of the Optical Society* 56, [1966]: 970.)

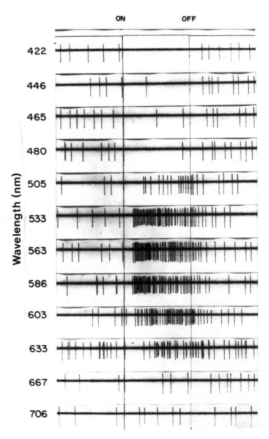

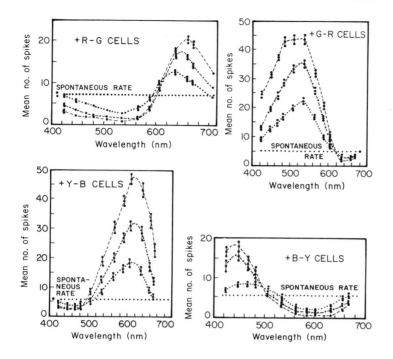

Fig. 3.12. Rate of firing of nerve cells during stimulation by light, plotted as a function of wavelength. These curves are averages, made for each of the four types of cells: +R−G cells show excitation for red light, inhibition for green; +G−R is its mirror image, excitation for green, inhibition for red; similarly, +Y−B and +B−Y for yellow and blue. Three curves are shown in each graph corresponding to different light intensities. Horizontal dotted lines indicate the firing rate in the absence of stimuli. Since animals are anesthetized, the firing rates are lower than they would be under normal conditions. (Adapted from R. L. De Valois, I. Abramov, and G. H. Jacobs, *Journal of the Optical Society* 56, [1966]: 972 and 973.)

called red, green, and blue. But at a deeper level in the nervous system, cells have been studied that respond in an antagonistic way to the pairs of colors red-green and blue-yellow.

When the fibers of the optic nerve are followed into the brain, they terminate in a small region known as the lateral geniculate nucleus (LGN). Information is transmitted through this region directly to the visual cortex. In the laboratory of Russel De Valois a series of measurements were made of the electrical responses of LGN cells of the macaque monkey. The animal is anesthetized and held rigid as an extremely fine electrode (diameter about 0.0001 cm) is inserted into the LGN region. The experimenter does not see which cell the probe approaches but samples many areas and obtains responses from many individual neurons. These cells, as discussed in the previous chapter, fire at a certain average rate, and their response to stimuli may be an increased rate of firing *(excitation)* or a decreased rate of firing *(inhibition)*. The subject's eye is stimulated with monochromatic light, and the response of the cell is measured as a function of wavelength (fig. 3.11). Some cells show excitation for all wavelengths, and others show inhibition for all wavelengths. But one class of cells responds with excitation for certain wavelengths and inhibition for others. Within this class, called *spectrally opponent cells,* four distinct types are found (fig. 3.12): one type shows maximum excitation in the red region of the spectrum and maximum inhibition in the green. A second type is a mirror image of the first, with inhibition in the red, excitation in the green. A third type shows maximum excitation in the yellow and maximum inhibition in the blue; the fourth type is the mirror image of the third. Each cell responds in one way or another to almost every wavelength in the spectrum; but the suggestion is clear that, at this stage in the neural pathway, the brain is being fed source data for the perception of four psychologically independent colors—red, yellow, green, and blue. Moreover, a mechanism exists whereby blue and yellow light may act in an opposing manner. In cells of type three or four, for example, some combination of blue and yellow will produce no response—effectively a "cancellation" of the two lights. Similarly, red and green are antagonistic in cell types one or two.

The spectrally nonopponent cells, those that respond in the same way for all wavelengths, carry information about light intensity, although it is known that opponent cells also respond to intensity. Recent data suggest that spectrally opponent cells respond to saturation, while nonopponent cells do not.

Research on processing of color information is still active, and many questions remain unanswered. The mechanism by which signals from the three photoreceptors are translated into the responses of the LGN cells is a subject of current theoretical and experimental study. Complex interactions exist between color and spatial location along the retina (an interaction related to simultaneous-contrast experiments) and among the three dimensions of color—hue, saturation, and brightness.

We began by remarking, in a subjective way, on the psychological reality, perhaps one should say, uniqueness, of yellow, along with red, green, and blue. A number of psychological experiments indicate that individuals feel a need to have these four color terms, at a minimum, to describe the spectrum and that they can relatively easily dispense with other terms. Moreover, although the set of color words varies widely from one language to another, certain intriguing consistencies are found. In a study of many different and historically unrelated languages, B. Berlin and P. Kay found that the basic color names appear in the same "order" among the languages. By order is meant that colors fall into categories which are (simplifying slightly) A: black and white; B: red; C: green and yellow; D: blue; E: brown; F: purple, pink, orange, and gray. If a language has only three color words, they will be black, white, and red. If a language has only five color words, they will be black, white, red, green, and yellow. A language will not have brown if it does not also have all colors in categories A, B, C, and D. Thus, after black and white (which have no hue and measure only brightness), humans seem to feel a need for names for the four major colors, the original Hering tetrad, red, yellow, green, and blue.

How do we know that when a speaker of Hungarian says *piros* he means the same "red" that we mean in English? Such questions can be answered by showing color charts to speakers of other languages. The chart contains an array of hundreds of colors varying in hue and brightness. If there is a term known to be roughly equivalent to our "red," the subjects are asked to locate the best representative of their "red" on the chart. Results for the four major colors are found to be quite consistent across languages.

We start with the naive assumption that the division of the spectrum is arbitrary. It is a continuum—indeed color is a three-dimensional continuum—and humans will apply color terms drawn from their distinct cultural and natural environments (the Eskimos have many words for white). However, although color terms do vary widely from language to language, linguistic and psychological data seem to converge with neurological experiments in suggesting that the four major colors are uniquely perceived across the human species.

The Electron and the Photon

4

There are three stages in the life of a light beam: it is created, it travels through space, and it is destroyed. The theory of electricity and magnetism, marking the high point of nineteenth-century physics, explained how light travels through space. Moreover, this theory can be formulated in a manner independent of the properties of matter, for light can travel through a vacuum. But light is created and destroyed only via its interaction with matter, from glowing gases of the sun to rhodopsin in the eye. Thus ultimately the theory of light could not be complete without an understanding of the fundamental nature of matter.

This understanding was finally achieved in the quantum theory of 1925, which provided for the first time a satisfactory explanation of how matter is constructed of atoms and molecules, how atoms are constructed of nuclei and electrons, and how atoms interact with light. Each of the major currents of nineteenth-century physical science played its part in leading up to the quantum theory. These currents were electromagnetic theory, the laws of chemical combination, the theory of gases, and statistical thermodynamics (a new theory of heat developed extensively by the Viennese physicist Ludwig Boltzmann).

Matter and Molecules

Chemistry came of age in the latter part of the eighteenth century when scientists began to make accurate measurements of the quantity, by weight, of materials undergoing chemical reactions. A crucial theoretical development took place in 1808 and 1810, when John Dalton, a self-educated English schoolteacher, showed that these measurements could be explained by a molecular hypothesis: that each element consists of many identical atoms, each compound of many identical molecules, and each molecule of a small number of atoms; and that a chemical reaction is simply the combination or recombination of several atoms to form molecules, the process multiplied manyfold to become observable in the test tube. A similar hypothesis for chemical reactions among gases was put forward in 1811 by the Italian Amedeo Avogadro.

The molecular hypothesis was also applied with great success to the physical behavior of gases. The assumption is that a gas consists of widely separated molecules (hence it is easily compressed), interacting very little with each other and moving incessantly at high speeds in a random manner. On this basis James Joule, in 1848, explained the pressure of a gas on the walls of its container. Other properties of gases, including their expansion when heated and their diffusion through air, were deduced from this *kinetic* molecular theory by various theorists in the mid-nineteenth century.

Finally, out of this work on gases, there developed the statistical theory of heat, perhaps the most subtle and yet the deepest of these nineteenth-century currents. Boltzmann argued that, although the molecules of a gas—or indeed of any matter—are so numerous that one can never predict the exact path of each one, nevertheless one can predict the behavior of the gas as a whole by giving a set of *probabilities* for each molecule's being in a particular state: for example, the probability that a molecule has a speed of 400 meters per second or the probability that a molecule is in the upper half of a container. Any system composed of many molecules, therefore, was to be studied by means of the laws of what came to be called statistical mechanics. These laws provided not only a firm foundation for the kinetic theory of gases but also an explanation for the phenomena of temperature and heat: temperature is a measure of the average speed of these random molecular motions, and the process of adding heat to a system is no more than supplying additional energy, thereby increasing these random speeds.

Boltzmann's work, which drew upon earlier work of Rudolf Clausius, Maxwell (again!), and others, was carried out over a number of years in the 1860s and 1870s. The molecular theory of matter thereafter gained increasing (although not universal) acceptance, with Boltzmann as its champion. As the new century approached, physics turned its attention more and more to the question of the structure of atoms and molecules themselves. And in this endeavor three different manifestations of light played key roles—leading first to paradox, then to resolution—in the emerging quantum theory of the atom. Thus, at the level of the very small, the story of light became intricately interwoven with the fundamental constitution of matter.

Blackbody Radiation

We see many objects by reflected light, but some objects are self-luminous: the sun, the stars, fire, the filament of an incandescent bulb. These sources of light have one thing in common: they are all hot. We imagine an experiment in which an iron bar is heated in a fire and then taken out of the flame; it glows red-hot—a self-luminous source. The temperature would be about 3,000 degrees Celsius. Heat it further, to about 6,000 degrees, and it glows white-hot. Now let it cool down in the air. The color gradually becomes pale red, then a deeper red, and finally luminosity vanishes (and you begin to see it, again, only by reflected light). At this point you can still feel heat radiating from the bar if you hold your hand a few inches from it. After a time this too stops.

Detailed study of this radiation in the nineteenth century led to the discovery of the following facts. The spectrum is continuous like the sun's, including infrared, visible of all colors, and ultraviolet. A graph of intensity versus wavelength shows maximum intensity at one wavelength and decreasing intensity at higher and lower wavelengths. The hotter the object is, the shorter will be the wavelength of a maximum intensity. The hotter the object is, the greater will be the total amount of radiation coming from a given surface area. For example, at 3,000 degrees maximum intensity occurs in the infrared, but there is enough intensity in the (visible) red for us to see it; the object is "red-hot." At a lower temperature the maximum is in the far-infrared. Now, although red is still the brightest of the visible wavelengths, it is all too dim for us to see. Our skin is sensitive to the infrared, which we feel as radiant heat. At 6,000 degrees the maximum wavelength is in the blue, with substantial radiation in the entire visible region. The distribution looks very much like that of the sun; that is, the variation of intensity with wavelength is similar to that of the sun, although the total radiation from the sun is of course much greater. The object is "white-hot." [1] The nature of this radiation depends somewhat on the kind of surface which is radiating, but the major features of the temperature dependence are rather uniform.

These relationships, involving the intensity and wavelength of emitted light and the temperature of the luminous surface, posed a great challenge to the physicists of the late nineteenth century. On the one hand, the science of statistical thermodynamics had just come into its own, providing a secure foundation for the concept of temperature and its connection with molecular motion. In the case of gases this motion consists of the free flight of molecules through largely empty space. But in the case of solids molecular motion was thought of as the oscillation of a molecule about a fixed position in a lattice. (Thermal motion in liquids is more complicated.) On the other hand, electromagnetic theory had predicted that oscillating electric charges would emit radiation. Electric charges in the antenna of Heinrich Hertz emitted long-wavelength radiation, and the suggestion being pursued now, by Max Planck (fig. 4.1) and other physicists in central Europe and England, was that the oscillation of individual molecules in a solid could be taken as the source of emitted light.

Fig. 4.1. Max Planck (Courtesy AIP, Niels Bohr Library)

The theoretical study of this problem marked one of the major turning points in the history of science, and for this reason we consider it in detail. Lacking knowledge of the nature of molecules (lacking even proof of their existence), physicists tried, by means of a series of indirect arguments, to formulate the problem in general terms, in terms that did not depend on the nature of the material emitting the light. First they introduced the concept of the *blackbody,* a body whose surface absorbs all radiation incident upon it (hence the term "black"), and proved that (1) any two blackbodies *at the same temperature* will emit precisely the same radiation and (2) a blackbody will emit more radiation than any other surface[2] at the same temperature. (see app. B.) A blackbody is an idealization, which may be approximated by such surfaces as black velvet or soft coal. But the best black surface is not a surface at all but a small hole admitting to a large cavity. (see app. B.) Think of the blackness of a hole punched in the top of a tin can, or the mouth of a cave, or the pupil of the eye (discount the highlights, which are due to specular reflection by the smooth surface of the cornea). If we build a large furnace, heat it to high temperature, and then allow radiation to emerge from a small hole in the furnace, we are producing blackbody radiation characterized solely by the temperature of the furnace.

The theoretical question is now the following: what sort of radiation (what intensity and what distribution of wavelengths) will be established in the interior of a cavity held at a given temperature? For this is the radiation that will stream out through the hole. Radiation is due to oscillating atoms in the walls of the cavity, but we have very little information about these atoms. For a blackbody, however, such information is irrelevant, for the answer depends only on the temperature.

These questions occupied the attention of many of the leading theoretical physicists during the 1890s. The first step was to demonstrate a relation between the energy of the oscillators in the walls and the radiation in the cavity. This relation was a proportionality between the quantity of radiation of a given frequency (and corresponding wavelength) and the energy of the oscillators having that frequency. Experience with oscillators in the everyday world (a pendulum, for example, or a violin string) shows that they normally have a characteristic frequency. A pendulum oscillates at its characteristic frequency, say, one swing per second in a clock, and keeps that frequency whether it is swinging in a wide or a narrow arc. Depending on how widely it swings, the oscillator of a given frequency can have various quantities of energy, from zero (if it is at rest) and upward. The key question in the theory of cavity radiation becomes, thus, what are the possible energies for an oscillator having a given frequency? Given an atomic oscillator of a certain frequency, what is the likelihood that it will have any particular energy, from zero and up?

These were just the kinds of questions that statistical thermodynamics was designed to answer. Boltzmann's theory had been constructed with reference to the behavior of gas molecules, but scientists concerned with radiation soon applied the theory to the oscillating molecules supposed to exist in the solid walls of a cavity. Work by Wilhelm Wien in the period around 1894–96 showed that statistical theories could account reasonably well for the measured intensities of blackbody radiation. But in the fall of 1900 new experimental results were obtained in Berlin by H. Rubens and F. Kurlbaum, which showed that radiation observed at longer wavelengths was in disagreement with Wien's theory. Max Planck, professor of physics at the University of Berlin, had been actively working on the theory of the radiation problem since 1897. He was a relatively late convert to the ideas of statistical thermodynamics but he had been using them in his most recent research. Working feverishly in the weeks following the experimental report, he discovered a way to modify the statistical theory to account for the data. Planck's modification was to assume, first, that each oscillator can have only one of a discrete set of energies rather than any energy above or equal to zero. If ε stands for a fundamental unit of energy, the oscillator can have only one of the energies 0, ε, 2ε, 3ε, and so forth. Second, the fundamental unit of energy for an oscillator is proportional to the frequency of the oscillator, $\varepsilon = h\nu$. The symbol ν is the Greek letter *nu*, representing frequency, and h is a constant, which has since come to be called *Planck's constant*.

Max Planck was an experienced and respected scientist and was not accustomed to taking radical steps. Yet his assumption was one of the most daring and momentous departures in the history of physics. It violates the principle of continuity, a principle so obvious and ingrained that it lies, unspoken, at the heart of classical physics. If I can travel at 5 mph or 10 mph, then surely I can travel at 7.5 mph or 8.75 mph or at any speed whatsoever in between. A pendulum may swing left to right over 2 cm, 3 cm, or with any amplitude over a continuous range. The oscillators in Planck's theory cannot. The principle of continuity seems to hold in everyday experience. (A significant exception is discussed in app. C.) Yet we can see with hindsight that it does not necessarily follow that it holds on the scale of atomic phenomena, where we have no experience. Planck's analysis of blackbody radiation was the first inkling that the atom would be different.

Planck's confidence in his theory stemmed not only from its agreement with experiment but from his awareness that he had dealt with a fundamental problem in a fundamental way; that is, in

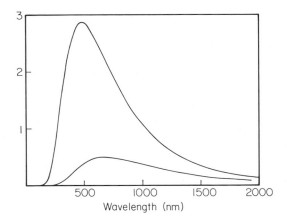

Fig. 4.2. Blackbody distribution curves according to the formula derived by Max Planck. Shown is a measure of the energy content of radiation plotted against wavelength. The upper curve is for a blackbody of temperature 5,800 °C. The lower curve is for 4,000 °C. At higher temperatures the total amount of radiation is greater at all wavelengths and the maximum intensity occurs at shorter wavelengths.

a way which did not depend on the detailed structure of the atom. He obtained, by comparing theory to experiment, a numerical value for the constant h, and he recognized that it would play a fundamental role in modern physics, bridging the gap between the atomic level and the macroscopic level (fig. 4.2). His paper, first presented to the German Physical Society in December 1900, was not immediately acclaimed; neither was it ignored. The theory was approached cautiously during the first few years of the new century, the feeling being that something lay hidden behind the new concept of ε, the elementary unit of energy (the *quantum,* as it came to be called).

The Photoelectric Effect

Gradually, in the decade from 1900 to 1910, the quantum theory began to be explored: by the Dutch physicist H. A. Lorentz (see fig. 10.5), something of an elder statesman in this period; by Lord Rayleigh (see fig. 5.2), then England's leading theorist; by Paul Ehrenfest, James Jeans, and Albert Einstein. In 1911 the first Solvay Congress was held in Brussels, attended by most of the prominent physicists of the day, and it became apparent at that time that the quantum could not be ignored. Among a number of developments bearing on the subject, probably the most important was Einstein's treatment, in 1905, of the *photoelectric effect.*

Here is the second phenomenon of light, which led finally to the modern theory of the atom. Heinrich Hertz, in the course of his experiments on transmission of electromagnetic radiation, observed that ultraviolet light could cause an electric current to flow across the air gap between two highly polished metal spheres. Other investigators quickly followed up this discovery. The photoelectric current persists even if the metal spheres, or metal plates, are placed in a vacuum, showing that the electric particles are not from the air. The current can be produced by longer-wavelength radiation when certain metals (generally those that are more active chemically) are used for the plates. With the alkali metals (for example, sodium and potassium) current can be produced with visible light. Through experiments with electrically charged metal plates it was determined that the particles carrying the photoelectric current were negatively charged. Shortly after the discovery of the electron in 1897, it was verified that the particles released in the photoelectric effect were electrons.

Subsequent developments underscored the importance of the wavelength dependence of the effect. For, as already implied above, in general relatively *short* wavelengths are most effective in producing the photoelectric current.[3] For each metal there is a maximum possible wavelength that will produce a current; with a longer wavelength no current is produced, no matter how intense the light is made. Furthermore, it became possible to measure the energy of the photoelectrons, and it was found that this energy also was independent of the intensity of light. These were the facts that remained inexplicable until 1905, when Einstein, then entirely unknown in the scientific community, presented what he called a "heuristic" view of the nature of light.

Einstein's starting point was not the photoelectric effect but the problem of radiation inside a cavity at high temperature. Without relying on Planck's hypothesis, he found that, at least in the case of short-wavelength (high frequency) radiation, there were certain similarities between the properties of electromagnetic radiation in an enclosure and the properties of a gas of molecules in an enclosure and that these properties could be understood if one thought of the radiation as consisting of particles, each of which had an energy equal to $h\nu$. (Again ν stands for the frequency, and h is Planck's constant.) Here Einstein goes a step beyond Planck's hypothesis, for Planck had quantized the atomic oscillators in the walls of the cavity. In retrospect it seems logical to assume that if the oscillators can have only one of a discrete set of possible energies, then when one gives up energy to create radiation, it will move down from one energy level to another, say from 5ε to 4ε, and therefore the light will have energy ε; in other words, light energy will also be quantized in units of ε, which is equal to $h\nu$. But physicists after Planck thought that after an atom gives up energy, this energy is in some way spread out over the large volume in which the light wave is situated and does not retain any quantized, or particle-like, character. Einstein proposed that light itself is quantized, not only in the act of its creation but in the free wave that travels across empty space.

In his 1905 paper he showed that the mysterious observations connected with the photoelectric effect could be easily explained on the basis of the concept of the quantum of light. For light of short wavelength (high frequency) contains energetic quanta; light of long wavelength contains less energetic quanta. If we suppose the electrons near the surface of a metal to be bound to the surface by a force of a certain strength, a more-energetic quantum may be able to break that bond and release the electron into the vacuum above the surface. A less-energetic quantum cannot, and so no photoelectric current is produced. Increasing the intensity of light simply produces a greater number of quanta, none of which is strong enough to free the electron.[4] The picture that evolves of the photoelectric effect is very much like a collision between two objects, say two balls on a billiard table. The difference is that after the collision between the quantum and the electron, one of them, the quantum, has disappeared. It has been absorbed, in the same way that light is always absorbed when it hits a black surface. But as a result we can say that all of the energy in the quantum is now possessed by the single electron with which it collided. In escaping from the metal the electron may lose some of that energy as it overcomes the forces that bind it. The electrons emerging in the photoelectric current are observed to have a range of energies; those least strongly bound originally in the metal (in some sense, those that were closer to the surface) have the maximum energy. This maximum energy is found to depend only on the frequency of the light. This again is consistent with the hypothesis that light deposits a fixed amount of energy on each liberated electron, the quantity of energy being proportional to the light's frequency.

Despite these successes, Einstein did not regard his paper as a definitive comment on the nature of light (unlike a second paper published in 1905, which propounded the theory of relativity and, in a stroke, resolved other paradoxes in the theory of electromagnetism). The wave properties of light had been firmly established by experiments beginning with Young and Fresnel and by the comprehensive theory of Maxwell. Einstein saw, therefore, that the discovery of particle-like properties in light presented a grave challenge to those seeking to understand the structure of physics; that there were inherent contradictions between Maxwell's electromagnetism and the accepted statistical view of oscillating molecules. He felt that profound changes in our understanding of light would soon have to be made, although he did not know how this was to be done.

Nevertheless, the quantum theory continued to have its successes. It explained certain unresolved problems having to do with the thermal properties of metals. Further work was done on the photoelectric effect. (In 1916 the American Robert A. Millikan succeeded in verifying Einstein's theory in detail. In particular, he determined a value for Planck's constant, h, that was in agreement with the value obtained from blackbody radiation.) The fundamental paradoxes of light remained, but beginning in 1911, a new line of research was set in motion—based on a third manifestation of light—giving new impetus to the quantum theory and leading eventually to resolution in the 1920s.

The Bright Line Spectrum

When objects are charged with static electricity a faint glow of light surrounds them. If you take clothes out of a dryer and into a dark room, you can easily see flashes of light as the clothes are separated. Static electricity is readily created by rubbing a glass tube; if such a tube is partially evacuated and filled with a gas at low pressure, you may observe a glow from the tube, with colors characteristic of the gas. Similar colors are seen in flames when certain volatile metals are vaporized in the flame: yellow for sodium, red for strontium, violet for potassium. Many observations of these kinds of luminescence were made in the 1700s and before by Otto von Guericke, John Bernoulli, Francis Hauksbee, and others. In 1752 Thomas Melvill, a Scot, studied the light of a flame passed through a prism and discovered that the spectrum was not continuous like that of the sun or a red-hot poker. Rather, parts of the spectrum were bright, and others were dark. This observation was made possible because he interposed a pinhole before the prism, forming, as had Newton, a narrow beam of light. Melvill found that the locations of the bright spots and their colors were different when different chemicals were placed in the flame.

The modern science of *spectroscopy* finds its beginning in this work and in the development of the *spectroscope* in the early nineteenth century by William Wollaston and Joseph Fraunhofer (fig. 4.3). The pinhole was replaced by a slit, a tube was added to create a still narrower beam, and the resulting spectrum, beyond the prism, was viewed through a telescope (fig. 4.4). Fraunhofer also invented the *diffraction grating*, to replace the prism. This was a series of parallel wires separated by about 0.005 cm, which acted like an enhanced version of Young's two-slit experiment. Light passing through the spaces between the wires forms an interference pattern beyond the grating, and the direction for constructive interference is different for each wavelength. Thus white light is spread, or dispersed, into the spectrum, and the degree of dispersion is increased by narrowing the slits. Modern diffraction gratings are made by ruling parallel lines on a glass plate, more than 10,000 lines per centimeter. A telescope views the grating at different angles and at each angle collects concentrated light of one wavelength. The grating spectroscope provided high-precision wavelength measurement, and this in turn laid the

Fig. 4.3. Joseph Fraunhofer (Courtesy AIP, Niels Bohr Library; W. F. Meggers Collection)

groundwork for the profound discoveries of twentieth-century atomic science.

Analyzed in the spectroscope, the spectra of flames and electric discharges in gases took the form of bright lines (see fig. 4.10), located at positions characteristic of particular wavelengths, all against a dark background. A unique set of wavelengths was associated with each chemical element, so that eventually the spectroscope became a device that could be used to identify the elements (indeed, very small quantities of the elements), just as fingerprints identify the individual.

Study of spectra of glowing gases was carried out systematically throughout the nineteenth century. In particular, the invention of reliable sources of electric voltage made possible the detailed study of luminescence from sparks and from the continuous flow of electric current through gases. Accurate measurements of the wavelengths of lines in these spectra were originally due to Julius Plücker, who built dumbbell-shaped glass discharge tubes with gas sealed inside and electrodes sealed into the ends. The narrow passageway between the two bulbs enhances the luminescence produced when voltage is applied to the electrodes, and the light is then studied through a spectroscope. Many investigators entered this new field, and during the latter part of the century many different gases were studied, and the wavelengths of their characteristic lines measured, in ultraviolet as well as in the visible. Some elements have thousands of lines. Sodium has a spectrum dominated by two lines in the yellow, so close together that they are often seen as one. Neon has strong lines in the red and orange. The neon advertising sign is just this kind of glass tube, containing neon gas through which an electric current flows. A bright yellow sodium spectrum is produced by the sodium

Fig. 4.4. Fraunhofer's spectroscope. The rectangular plate in the center is a diffraction grating. The telescope on the left can be moved around the circular platform, and a precise angular scale is ruled around the platform's circumference. (Courtesy AIP, Niels Bohr Library)

Fig. 4.5. J. J. Thomson (*left*) and Ernest Rutherford (Photo by D. Schoenberg; courtesy AIP, Niels Bohr Library; Bainbridge Collection)

vapor lamps that have made their appearance recently on highways and city streets.

The body of data on bright line spectra of gases posed a challenge to theorists as great as that of blackbody radiation; the challenge was in some ways greater, because the continuous emission from hot solid objects was largely independent of the nature of the object, whereas the line spectrum was different for each element. On the other hand, a gas is presumably the simplest form of matter, since each atom or molecule is independent, not affected by forces due to the other atoms. Furthermore, one can concentrate attention on a single chemical element, perhaps one with a simple structure, and hope to determine general principles which apply to the other elements. The simple structure that came to play this crucial part in the story of light and matter was the hydrogen atom. Chemical experiments and the observations of Dalton, Avogadro, the Frenchman Joseph-Louis Gay-Lussac, and others had already pointed to hydrogen as the simplest of all atoms. In the years 1909 to 1912 the first successful theory of the hydrogen atom, and its line spectrum, was formulated, based on the experimental genius of Ernest Rutherford (fig. 4.5) and the daring of Niels Bohr.

The Nuclear Model of the Atom

Rutherford's early work had been on the subject of radioactivity, which had been discovered in the closing years of the nineteenth century. He had shown that of the three types of radioactivity

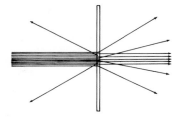

Fig. 4.6. Rutherford's experiment, the scattering of a beam of alpha particles by a thin metal foil

(termed alpha, beta, and gamma), alpha radioactivity consisted of particles with positive electric charge (now known to be helium nuclei). These alpha rays cannot penetrate a thick slab of metal; but if a stream of the particles strikes a thin metal foil, most of them pass directly through, and many of them are bent through small angles, forming a narrow spray, something like the action of the nozzle of a garden hose. Rutherford reasoned that this phenomenon, which subsequently became known as "scattering," might be used as a tool to study the nature of the atoms in the metal foil. For it was a collision between the alpha ray particle and an atom in the foil that caused the path of the particle to be deflected. (Rutherford's invention, the scattering experiment, remains to this day the physicist's most productive source of information about the atomic and subatomic world.)

Rutherford and his colleagues at Manchester University could observe the passage of individual alpha particles by using a fluorescent screen, made of a material that produces a scintillation, or a brief flash of light, in response to a charged particle. For long hours they sat in darkened rooms, first allowing their eyes to become dark-adapted and then observing the screen through a microscope and counting scintillations. (In modern experiments scintillation counters are still used to detect particles, but the flashes of light are counted electronically.) What they found was that indeed most of the alpha particles came directly through the foil, deflected not at all or only through small angles. But some particles were deflected through large angles, some practically reversing direction (fig. 4.6). Rutherford realized at once that this result was quite inexplicable under the then current view of the structure of matter. As he later wrote: "It was quite the most incredible event that has ever happened to me in my life. It was almost as incredible as if you fired a 15-inch shell at a piece of tissue paper and it came back and hit you."[5] Although atoms were not understood in detail, their sizes and masses were known roughly (about 100 million lined up would reach a centimeter), and it was supposed that in a solid, like the gold foil that Rutherford used as a scatterer, the atoms lay in contact with their neighbors, like oranges in a crate. The gold foil was very thin, but an alpha particle would still have to pass through several thousand layers of atoms before it could penetrate. Theory showed that with so many scatterings, there would be roughly an equal number to the left and to the right, so that on the average deflections would cancel each other; there was no way in which they could add up to a deflection of 90 degrees or more.[6]

Taking the opposite tack, Rutherford considered the possibility of an alpha particle being scattered backward by a collision with a *single* gold atom. The atom contains electrons; that much was generally accepted at the time based on the appearance of electrons in a variety of experiments with different materials. But the electron could not play a role in scattering alpha particles, since its mass is very small, about $1/7,500$ of the mass of the alpha particle. In a collision between the two only the electron recoils; the alpha particle is scarcely deflected. The rest of the atom, possessing most of its mass, consisted of some unknown background material, positively charged in order to balance the electron's negative charge (to form electrically neutral matter). What is to be expected if the alpha particle is scattered by the positive constituent of a single atom?

Two positive charges repel, so if the alpha particle passes near the target, it will veer to the side; if it approaches the target almost head-on, the repulsive force causes the alpha to recoil in an almost backward direction. Most important, the alpha approaches the positive constituent of a gold atom to within a distance of about 10^{-14} meters (according to calculations Rutherford was able to make using the known value of the alpha's energy). This distance is to be compared with the typical dimension of the atom, around 10^{-10} meters. Small as the atom is, the positive component in it is more than ten thousand times smaller.

Thus was born the nuclear, or planetary, model of the atom. The positive component is a *nucleus,* occupying a minute volume at the center of the atom, though containing most of the atom's mass. In Rutherford's experiment most alphas go right through the wide empty spaces between nuclei and are not scattered at all. The few that by chance come near a nucleus are scattered in various directions. The empty space is not truly empty, for it is occupied by electrons. There is an attractive electrical force between each electron and the nucleus, and this force is what binds the atom together. Finally, the electron cannot be stationary, for if it were, it would fall into the nucleus. Rather, just as the earth orbits the sun under the influence of the attractive force of gravity, the electron orbits the nucleus. We are thus led by Rutherford's work to a planetary model of the atom.

Into this revolutionary situation stepped the young Danish theorist Niels Bohr (fig. 4.7). He saw, as Rutherford did also, that the planetary model, while an elegant explanation of the scattering experiments, nonetheless did great violence to the classical Maxwell theory of radiation. For according to that theory a charged electron in circular motion will continually emit radiation and thereby lose energy. Its orbit will shrink and it will quickly fall into the nucleus. Bold steps were needed, and Bohr knew of the bold steps that had already been taken in the theory of light by Planck, Einstein, and the other theorists of the quantum. The world of the atom was different from the world of everyday experience, and the latter could not always be used as a guide. Bohr asserted that an electron in an atomic orbit simply did not radiate;

Fig. 4.7. Werner Heisenberg (*left*) and Niels Bohr (Courtesy AIP, Niels Bohr Library)

that at this level the theory of Maxwell did not apply. Rather, like the oscillators that produce blackbody radiation, the electron can exist in only certain *quantized energy levels;* and as long as it stays in a given level it produces no light, and its energy does not change. But an electron can spontaneously fall from one energy level, labeled, for example, 2, to a lower level, labeled 1; and when it does it loses an amount of energy that is the difference, $E_2 - E_1$. This energy is lost by the atom, and it goes to form, to create, a quantum of light. Thus light is the child of the atom in the act of change and, in another sense, is the mechanism of the atomic process; for if there were no light, there would be no way for the atom to go from one state to another.

Bohr directed his attention to the hydrogen atom, since it is the simplest possible atomic system: one proton orbited by one electron, the two held together by electrical attraction. He proposed a quantization condition determining the possible circular orbits for the electrons and deduced from it the set of possible values, $E_1, E_2, E_3, \ldots$ for the energy of the atom. This sequence, although not evenly spaced as in Planck's oscillator, nevertheless determines the possible frequencies of light that can be emitted by the atom. Bohr's final step was to adapt the Einstein equation for the frequency of light emitted, introducing Planck's constant and asserting that the frequency of emitted light will be given by $h\nu = E_2 - E_1$ when the atom makes a transition from state 2 to 1. The frequencies, hence the wavelengths, of emitted light are determined, and since the energies and the energy differences are quantized, so are the frequencies. The isolated atom emits light of only certain definite frequencies, as observed in the bright line spectrum. Moreover, these frequencies fall into categories: transitions from state 2 to state 1, from state 3 to state 1, from 4 to 1, 5 to 1, etc., form the first series; transitions from state 3 to 2, from 4 to 2, 5 to 2, etc., form the second series; and so on (fig. 4.8). In Bohr's time, lines of the first three series had been observed, and more series have been seen since.

Each of the possible states of the atom corresponds to a circular orbit of fixed radius, the lowest energy state having the smallest radius. (Later developments allowed orbits in the shape of an ellipse). Atoms normally stay in the lowest energy state, the *ground* state; but for various reasons they may become "excited" and raised to one of the higher energy levels. In a gas discharge tube excitation is due to the electric voltage applied at the ends of the tube; in the sun it is due to collisions among the atoms. Whatever the mode of excitation, the atom later spontaneously "falls" down, either to the ground state or to one of the intermediate states and then in several steps down to the ground state. In each of these steps the characteristic spectral light of hydrogen is emitted.

The equations of Bohr's theory are in very precise agreement with the observed frequencies in the hydrogen spectrum. The observations can also be used backwards, to determine the value of Planck's constant. The value arrived at in this way was found to be in agreement with the result deduced from blackbody radiation and in agreement with the result deduced (several years after Bohr's paper) from measurement of the photoelectric effect.

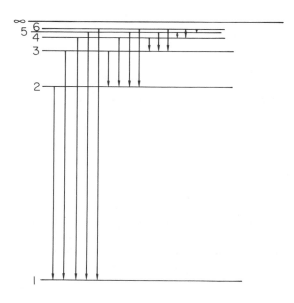

Fig. 4.8. Energy levels of the hydrogen atom according to the Bohr theory. The first six levels are shown and drawn to scale. Level 1 is the atom's ground state. Light is emitted whenever the atom makes a transition from a higher state to a lower one, and the frequency of light is proportional to the energy difference. Higher levels are closer together. The line labeled infinity (∞) represents the energy the electron would have if it became barely free; that is, just able to escape from the nucleus.

Looked at in terms of the electron's orbit, the measurements determine the radius of the atom in its ground state—what one might call the "size" of the hydrogen atom. Thus we may answer the question, Why is the atom so small? Indeed the size of an atom, quite generally, is determined by three things: Planck's constant, the mass of the electron, and the electric charges on the electron and the nucleus. It is not that the atom is small: it is only small compared with us. The real question is, why are we so large, or why are we made of so many atoms? And that, of course, is because we are such a complicated piece of machinery.

Elaborations of the Bohr Theory

One way to excite an atom is to use light as a *source* of energy. Shine a light beam into a gas, and let the frequency of the light be such that the quantum energy ($h\nu$) is just equal to the difference in energy between the ground state and some higher state of the atom. Then the quantum will be absorbed and the atom will jump up into that higher state: the reverse process of atomic emission of light. If the light beam is not monochromatic but has a continuous distribution of frequencies, then just those frequencies (energies) that correspond to transitions of that atom will be absorbed; the rest of the beam passes through the gas unchanged. The spectrum of transmitted light will thus contain *dark* lines at just those frequencies that are absorbed; these are called *absorption lines*.

Fraunhofer, in his researches with the spectroscope, had discovered such dark lines in the spectrum of the sun (fig. 4.9). (Now one can identify thousands of lines.) Later investigators found similar dark lines in the spectra of stars, although not in the spectra of terrestrial blackbodies. The key to understanding absorption lines came when it was discovered that the frequencies at which they occurred were the same as the frequencies in the *bright line* spectra of glowing gases (fig. 4.10). Gustav Kirchhoff, in 1859, showed that he could produce dark lines in the spectrum of a blackbody by passing that light through a region of cool gas. Furthermore each gas produced absorption lines at just the frequencies of the bright line spectrum of that gas. He deduced that such a process was responsible for the absorption spectrum of the sun: that the sun is surrounded by an atmosphere of relatively cool gases and that each gas absorbs its characteristic light, producing dark lines. In Bohr's quantized atom we see the simple explanation of this process. The energy levels are characteristic of each element. In an excited gas, as in a discharge tube, the atoms are in higher levels and fall down to lower levels, emitting light. In the atmosphere of the sun and stars, the same kinds of atoms absorb light in the process of going from lower to higher levels.

Here was a tool of profound importance to modern science, particularly astronomy, for the frequencies of spectral lines are a signature of an element. We determine that the sun contains vapors of sodium, calcium, iron, and many other elements, and we can probe the chemical composition of stars many light-years away. The element helium was first discovered via a series of unexplained dark lines in the sun's spectrum, lines which were not related to emission lines then known on earth. Later helium gas was found on the earth.

In the years from 1912 to 1925 the Bohr theory of the atom provided a conceptual framework for the study of matter on many fronts. Most important, the theory had to predict energy levels for atoms more complicated than hydrogen: helium with two electrons, lithium with three, and so on; thereby it would predict the

Fig. 4.9. A portion of the spectrum of the sun showing many dark lines. The series of film strips covers the wavelength region from 660 nm to 720 nm, in the infrared. (The film is labeled in angstrom units. One angstrom equals 0.1 nm.) (Courtesy Palomar Observatory)

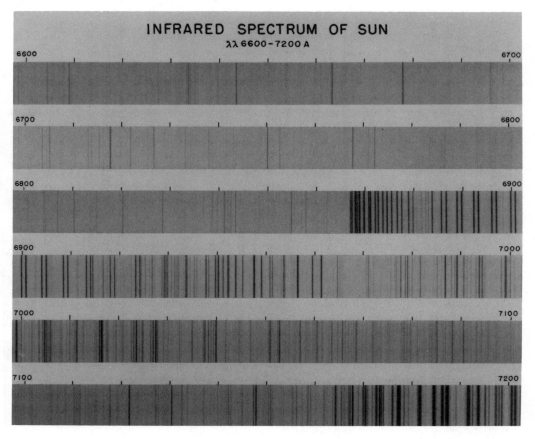

frequencies of emission spectra for these elements. In this effort the concept grew of electronic shells about the nucleus, the outer shells less tightly bound. In a many-electron atom only the electrons in the outermost shells are "loose" enough to be excited to higher levels by mechanisms such as the electric discharge through a gas. These outer electrons thus determine the element's spectrum.

In these years, too, the notion of electronic shells initiated the modern atomic theory of chemical reactions. The electrons in the outer shells, transferring from one atom to another or being shared by two atoms, provide the machinery for chemical reactions and the binding of atoms to form molecules. Finally, the Bohr theory provided an understanding of the properties of x-rays, which were being systematically studied in the years 1913 and 1914 by the English physicist H. G. J. Moseley.

X-rays are a form of electromagnetic radiation, produced when a high-energy beam of electrons strikes a metal. The wavelengths of this radiation are distributed over a continuum, on which is superimposed a series of spectral lines; typical x-ray

Fig. 4.10. The bright line spectrum of the vaporized element *iron* is shown with the spectrum of the sun. The wavelength region is from 300 nm to 330 nm, in the ultraviolet. The solar spectrum is in the center of each strip, and the iron spectrum is above and below it. The bright lines of iron occur at the same wavelengths as some of the dark lines in the solar spectrum. (Courtesy Palomar Observatory)

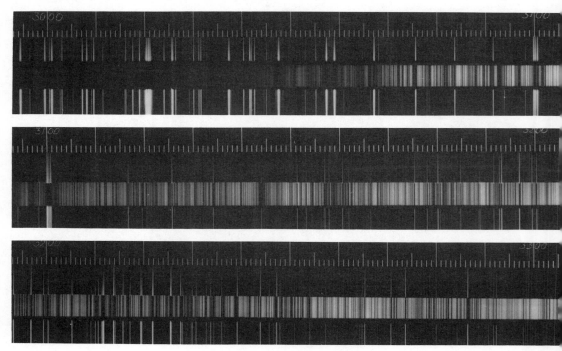

wavelengths are around 10^{-10} meters, or 0.1 nm (see fig. 1.28). The work of Moseley and Bohr showed that x-rays are due to transitions of the innermost electrons in a many-electron atom. The high-energy beam may knock one of the electrons in the inner shell completely out of the atom, leaving a vacancy in that shell. The vacancy is filled when an electron from perhaps the second inner shell falls inward, in the process emitting radiation. The difference between this event and the production of optical spectra is that the energies involved in transitions between inner shells are much larger, and the corresponding x-ray wavelengths are several thousand times smaller than those of light.

Just as the wave properties of light, such as diffraction, are difficult to observe because the wavelength is small on the scale of everyday objects, so the wave properties of x-rays are correspondingly more difficult to observe. Diffraction of light is observed with a diffraction grating ruled with closely spaced lines. X-ray diffraction may be observed by using a crystal as a natural grating. In the crystal atoms are arranged in a geometrical latticework, and adjacent rows of atoms play the role of ruled lines in a grating, separated by distances of around 10^{-10} meters. This is the typical size of an atom and also a typical x-ray wavelength. It was just such a diffraction experiment, in 1912, that established that x-rays were a form of electromagnetic radiation and determined their wavelengths. Today the situation is reversed, and x-ray diffraction is used as a tool, one of the most powerful tools, to study the structure of crystals.

Despite its successes, the Bohr theory did not emerge as the final theory of light and matter. (Nor do we expect that there will be a "final" theory.) On the one hand, although it provided a general framework for the discussion of atoms more complicated than hydrogen, it was not fully successful in predicting frequencies of spectral lines. More important, it did not resolve the fundamental issues that underlay the quantum theory of Planck and Einstein. Why are the states of oscillators and atoms quantized? What happens when a system changes its state and a unit of light energy is created? How can this quantum (later the word *photon* was proposed) behave like a particle in some cases and yet produce the interference patterns seen by Thomas Young?

Quantum theory

The answers to these questions, not all of which can be presented here, were formulated finally in the years 1925 and 1926, in the revolution of modern quantum theory, a revolution that shook the foundations of physics and philosophy. In a way, the central theme continued to be the nature of light, and what came to be called the *wave-particle duality*. But there were broader implications in the

Fig. 4.11. Louis de Broglie
(Burndy Library; courtesy AIP,
Niels Bohr Library)

new theory, and the first hint of these was given in 1924 by Louis de Broglie (fig. 4.11) in his doctoral dissertation. De Broglie proposed that objects that we ordinarily think of as particles may also have wavelike properties and that these wavelike properties would manifest themselves only in phenomena occurring on an atomic scale, as dictated by Planck's constant. The wavelength of these matter waves, for a given particle like an electron or proton, would be inversely proportional to the particle's mass and speed. Experimental confirmation of de Broglie's hypothesis, for electrons, was soon forthcoming (in 1927), but in the meantime theorists Erwin Schrödinger, Werner Heisenberg (see fig. 4.7), Max Born, and others developed these ideas into a new and fundamental theory.

It was ultimately a theory of measurement. Observations made by scientists are not made directly on the objects they are studying (a molecule, a virus, or a living cell) but rather on the medium of measurement. We look at a cell with light, and what we observe is the light reflected from it. It follows that the light disturbs the cell—changes it to a new state. If one measurement of a system disturbs it, then, if a second measurement is made, it is not being made on the same system as was the first measurement. Thus, in general, two measured quantities will be incompatible; that is, one cannot know the value of one measured quantity without there being created some uncertainty in the value of a second quantity. This limitation on our knowledge of the physical world, embodied in the Heisenberg uncertainty principle, is intimately associated with quantization. For if light were not quantized, we could imagine making measurements with smaller and smaller quantities of light energy, until we would ultimately approach a measurement that would not disturb the system being measured; or at least we could make the disturbance as small as we wished, so that the uncertainty principle could be reduced to negligible proportions. But nature presents us with h, Planck's constant, which determines the minimum energy content of a photon, producing uncertainties that cannot be made negligible and that are significant on the atomic scale.[7]

Moreover, modern quantum theory tells us what we *can* measure within the limits of these uncertainties. If one quantity is measured, then a precise value cannot be given to a second quantity. But the possible values of the second quantity are given probabilities, which can be calculated exactly from the equations of the theory. For example, suppose we know that the electron in the hydrogen atom is in its ground state; that is, we know its energy precisely. Then we cannot know other quantities, such as the electron's position at a given time or its velocity. But quantum theory does give us a means to calculate the probability of the electron's being at any particular point in space. Here we begin to see the

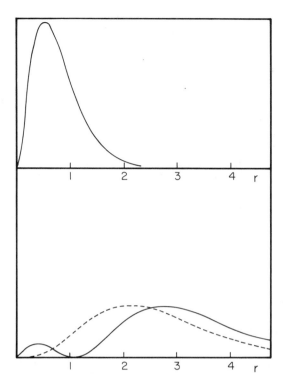

Fig. 4.12. Probability distribution for the electron in various states in the hydrogen atom. Plotted is the probability (in arbitrary units) of the electron being found at a distance, r, from the nucleus. The horizontal axis is r in units of 10^{-10} meters. The upper curve shows the case of the ground state. The probability of finding the electron at $r = 0$ (that is, inside the nucleus) is zero; the probability of finding the electron at large distances is small. The probability is maximum in the region around 0.5×10^{-10} meters, which is the radius of the orbit in Bohr's theory. The lower curves show the first excited state. Two curves are shown because there are four states that have the same energy, the second lowest energy for the hydrogen atom. Three of them have exactly the form of the dashed curve, and the fourth state has the form of the solid curve. Note that the probability here is concentrated at larger distances than for the ground state; one may say roughly that the "orbit" is larger. Note the characteristic wavelike structure. The phenomenon in which several states have the same energy is called *degeneracy*, and is found widely in atomic physics. Under certain conditions the degeneracy of the hydrogen atom is broken; that is, the previously degenerate states now have slightly different energies. This happens when the atom is placed in a magnetic or an electric field. These energy differences show up in the line spectrum as a splitting of a single line into several closely spaced lines. The presence of magnetic fields in sunspots is deduced from the splitting of spectral lines in light coming from those regions of the sun.

concept of a wave emerging. For a wave is something distributed through space, while a particle is thought of as located at a point. The electron in the hydrogen atom is not located at a point, but it has a probability distribution in space (fig. 4.12).

The Bohr concept of an electronic orbit is no longer tenable, but the new theory substitutes for the orbit a probability distribution. In the case of the hydrogen atom the probability is concentrated in the region where the Bohr orbit was. For a free electron (one not bound to the hydrogen nucleus) the probability distribution looks more like what we think of as a wave; it is a confined region of oscillations, called a wave packet. These wavelike properties are difficult to observe because typical wavelengths of electrons that might be studied in the laboratory (for example, the cathode rays in a television tube or an oscilloscope) are extremely short—around 10^{-9} or 10^{-10} meters. Wave effects like diffraction and interference are seen only when the typical dimensions of the laboratory system are not much larger than the wavelength. For light a thin slit 10^{-4} meters (0.1 millimeters) wide produces a clear diffraction pattern. For an electron the slit must be much smaller.

In 1927 C. H. Davisson and L. H. Germer (fig. 4.13) dem-

onstrated wave properties in a beam of electrons by using the latticework of atoms in a crystal as the equivalent of a diffraction
grating (just as had been done fifteen years earlier for x-rays).
Observing the electrons scattered by a nickel crystal, they found a
pattern of alternating minima and maxima, just like the pattern of
constructive and destructive interference seen in light passed
through a grating. Using the well-established equations of diffraction, Davisson and Germer calculated the wavelength appropriate
to the electrons in their beam and found a value in agreement with
the theory of de Broglie. Electron waves were soon confirmed by
other workers, including G. P. Thomson (the son of J. J. Thomson
[see fig. 4.5], who in 1897 had discovered the electron by observing its particle-like properties).

In an electron microscope a beam of electrons is generated
with wavelength much shorter than the wavelength of light. Thus
it is effective in resolving structures (viruses, for example) much
smaller than can be resolved by a light microscope. X-rays, of
course, have similarly short wavelengths, but the electron beam
has the advantage that, since electrons are electrically charged,
the beam can be manipulated by electric and magnetic fields—
indeed focused, just as light is focused by a lens.

Fig. 4.13. *Left to right:* C. H. Davisson, L. H. Germer, and C. J. Calbick. (Bell Laboratories; courtesy AIP, Niels
Bohr Library)

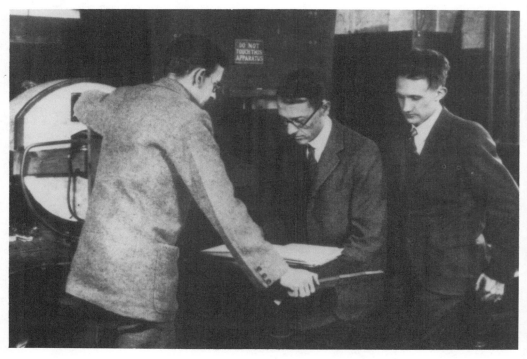

The quantum theory of 1925 resolves the dilemma of light by unifying the entities of light and matter and by asserting that wave-particle duality applies to both. It failed to address the issue of another fundamental distinction between the two: the photon is created out of nothing—in the flash of an atomic jump—while the electron is permanent. We return to this question in chapter 10.

The message of light in the twentieth century is the message of the atom, allowing a new understanding of how man himself interacts with his surroundings. We cannot dwell on the philosophical implications of the quantum theory, which extend to epistemology and logic, or on its predictive success in fields such as astrophysics, chemistry, metallurgy, nuclear and elementary particle physics, materials science, as well as atomic physics and optics. Suffice it to say that it remains today the indispensable tool of the theorist in physical science.

In the Atmosphere

5

Walk out on a fresh clear day in autumn and look at the deep blue of the sky. As your eye moves closer to the direction of the sun the sky becomes lighter, its color brightening to white and then to gold. The yellow-gold of the sun itself dazzles your eye and seems to send flashes of light across your field of vision. Your eye is involuntarily closed or averted, to rest calmly on the sky again. Closer to the horizon the blue of the sky becomes paler again— not dimmer, but closer to white.

In 1869 John Tyndall (fig. 5.1) demonstrated in his laboratory at the Royal Institution that the blue of the sky could be produced by the scattering of light by small particles. He used fine particles of dust suspended in the air inside a glass tube. With

Fig. 5.1. John Tyndall (*standing*). Faraday is seated on the left. (Courtesy AIP, Niels Bohr Library, Zeleny Collection)

a bright light beam aimed down the length of the tube, he was able to observe faint blue light coming sideways out of the tube. He reasoned that the blue sky was produced similarly by particles of dust in the atmosphere. But suppose we go to a mountain range, where the air is purer, and where the light's path through the atmosphere is shorter. Here, although there is less dust, the blue of the sky is not less but richer and deeper. It was Baron Rayleigh (fig. 5.2), the successor of James Clerk Maxwell as Cavendish Professor at Cambridge University, who first proposed that the blue of the sky is due not to dust particles but to the gases of the atmosphere, in other words, to the scattering of light by individual molecules.

Fig. 5.2. Lord Rayleigh (*left*) with Lord Kelvin (Photograph by A. G. Webster; from Robert John Strutt, Baron Rayleigh, *John William Strutt, Third Baron Rayleigh,* [Longmans, Green, and Co.]; courtesy AIP.)

The Earth's Atmosphere

The atmosphere is in fact a complex mixture of gases, vapors, crystals of snow and ice, droplets of water, and small particles of solid matter (dust), both natural and man-made. Its composition varies with the days, with the seasons, and over longer time cycles. The primitive atmosphere, when the earth was young, was very different from that of today, although the question of its composition is still unsettled and controversial. The original solar nebula—the cloud out of which the sun and the planets formed—was made predominantly of hydrogen and helium, with small amounts of carbon monoxide, methane (a compound of carbon and hydrogen), water vapor, and other gases. Clues to its composition are drawn from studies of the sun's atmosphere and also from the analysis of meteorites, which are believed to have formed directly from the original nebula. If gases of the solar nebula formed a primitive atmosphere around a newly aggregated solid earth, it is most likely that they did not remain. (This follows from the observation that the present atmosphere has a very low concentration, compared with the sun, of inert gases such as neon and krypton, elements which could not have been depleted by chemical reactions.) On the other hand, gases such as carbon compounds and water vapor could have been chemically incorporated in solid minerals (as they are in meteorites) and thus be trapped in the earth's interior. Earth's atmosphere evolved, according to the most widely held theory, by the process of "degassing," the release of gases from the interior via volcanic action. Whether this process occurred once and suddenly, soon after the planet's birth, or has been happening gradually and perhaps continually is an open question. It is certain, however, that volcanic activity was more vigorous in early geological times than it is now. Analysis of gases released by present-day volcanoes shows the presence of hydrogen, carbon monoxide, and various sulfur compounds, among other things. Similar compounds have also been found on Jupiter, which probably retained its early atmosphere.

The present atmosphere on earth consists primarily of nitrogen (78 percent) and oxygen (21 percent). The next most abundant constituents are water vapor and the inert gas argon. Argon is produced by the decay of a radioactive isotope of potassium, present in small quantities in the minerals of the solid earth. Oxygen, which is so essential to human life, was not present at all in the pre-biotic atmosphere. Indeed, under the conditions prevailing at those times, any free oxygen temporarily released would have immediately undergone some chemical reaction. It is now the generally accepted view that oxygen entered the atmosphere as a consequence of the evolution of life.

Experiments have shown that if a mixture of gases like that in the early atmosphere is subjected to electrical discharges (like lightning) or ultraviolet light, small quantities of amino acids, the building blocks of living protoplasm, are formed. Even some elementary proteins have been found in these studies, although nothing like the complex proteins and DNA molecules in living matter has been synthesized. But life, of course, exists, and the fossil record shows that there were living microorganisms as long as 3.5 billion years ago. Further, it is known that more than 3 billion years ago photosynthetic algae, which would take in carbon dioxide from the atmosphere and release oxygen, had evolved. This, we believe, is the process by which oxygen formed in our atmosphere, leading subsequently to the evolution of oxygen-consuming animals. The earth's formation is dated at only 4.6 billion years ago, and several hundred million years would have passed before the solidification of the surface rocks. So it is clear that the earth and living things evolved together. In other words, since the earliest times the earth has not been simply a stage upon which life developed but rather an interacting partner with our primitive ancestors, as it is today with us.

For example, there is in the stratosphere a layer of *ozone,* a gas formed by the combination of three oxygen atoms (in contrast to the ordinary oxygen molecule, which we breathe, made of two atoms). This ozone layer, produced by the action of sunlight on oxygen, now absorbs much of the sun's ultraviolet radiation. It thereby protects us from dosages of this radiation which would be ultimately lethal. Ironically, the ozone layer was not present in the early atmosphere, and the intense ultraviolet reaching earth's surface may have been instrumental in catalyzing the early chemical events that led to original primitive life-forms.

The chemical structure of the atmosphere, the seas, and the soil has been intimately connected with the growth of life on the planet since a time almost as far back as the origin of the solar system itself. In these interactions there has been a balance, what scientists call equilibrium, in which opposing forces counteract each other so that a complex system does not change, or changes so slowly that each element has time to adjust and survive. Only in the last million years or so has a life-form evolved sophisticated enough to disturb the equilibrium of the earth.

Through earth's atmosphere comes light from the sun and the stars, and in John Tyndall's laboratory it had been demonstrated that small particles would scatter light in all directions. Moreover the strength of this scattering depended on the wavelength of light, the shorter wavelengths—toward the blue end of the spectrum—being scattered preferentially.

Rayleigh's accomplishment, in work beginning in 1871, was to develop the theory of scattering of waves. He envisaged a rapidly oscillating wave of electric and magnetic fields moving across a molecule and setting up a sympathetic vibration in that molecule

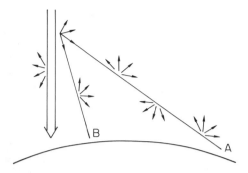

Fig. 5.3. Sunlight is scattered by air molecules as it passes down through the atmosphere. Since the shorter wavelengths are scattered more effectively than the longer, blue is removed from the direct beam of the sun. We see this scattered light when we look in directions that are not directly toward the sun, for example, in the directions along the solid lines from points A and B. But skylight produced high in the atmosphere must also pass through air molecules, and itself is scattered. Looking in the direction from A, closer to the horizon, the scattering of skylight occurs more frequently, and blue tends to be removed from this skylight. The sky looks blue-white or white. In the direction from B the secondary scattering has less effect because the path length is shorter. We see a more saturated blue.

at the same frequency as the incoming wave. This oscillating molecule then, in accordance with the recently published theory of Maxwell, emits radiation in all directions. It is a scattering process and bears a marked similarity to the scattering of a beam of particles as in Rutherford's work. In the case of a light beam, Rayleigh showed that scattering is all the more effective if the oscillation induced in air molecules is faster. Within the sun's spectrum scattering is therefore more effective for the blue light, which has the highest frequency of oscillation. Thus, in any direction we look, our eyes are struck by rays of light from atmospheric molecules previously stimulated by the rays of the sun. The spectrum of the sky contains all colors just as the sun's spectrum does. But it is weighted more heavily toward the blue end and so produces the sensation that we call sky blue.[1]

As we look toward the horizon the blue of the sky becomes less and less saturated, and low over the horizon the sky is always close to white. Blue light scattered by atmospheric molecules must itself pass through part of the atmosphere, which tends to remove the blue component of this skylight. If the path through the atmosphere is long, as is the case when we look toward the horizon, this secondary scattering becomes more important. The skylight then tends to lose its predominant hue and looks white (fig. 5.3).

Water, Dust, and Heat

Rayleigh's wave theory works for light passing through particles whose dimensions are much smaller than the wavelength of the light: molecules, for example, which are typically several thousand times smaller than wavelengths of sunlight. On the other hand, dust particles and water droplets are often larger than these wavelengths. Sunlight striking such objects is primarily reflected, producing scattered light that is white, like sunlight itself. On a dry summer's day the sky looks hazy. The lack of rain has left an accumulation of dust in the air and the sky appears white. Simi-

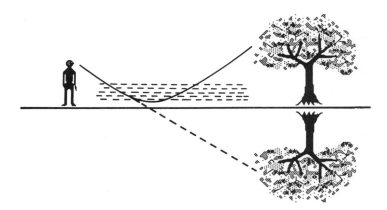

Fig. 5.4. Curvature of light rays in a layer of warm air close to the ground. A mirage is seen in the form of inverted objects below the surface, suggesting a layer of water on the surface.

larly mist in the air, or a high level of pollution, makes the sky appear white or gray.

The brain interprets light beams that enter the eye from a given direction as emanating from objects out in that direction. Scattered blue light is produced by the atmosphere everywhere along the line of sight. So a distant object, a town or a mountain, may appear hazy and bluish, because superimposed upon the light beams from the object are the blue light beams from the intervening atmosphere. The earth itself, viewed from an airplane at 30,000 or 35,000 feet, has a blue cast. From a rocket beyond the atmosphere we see very little of the surface of our planet; looking down we see the clouds and the blue of the sky.

But on the moon there is no sky. The moon has no atmosphere; there is no scattered light; and when the astronaut looks in a direction away from the sun, he sees only the stars and the dark. Moreover, the sun looks pure white. The sun we see at the bottom of earth's atmosphere appears yellow because the shorter wavelengths have been scattered out of its direct beam. The resulting spectrum is weighted more toward the longer-wavelength end; we see a desaturated yellow.

Fine water droplets in mist and in clouds scatter light equally for all colors but do not scatter equally in all directions. Rather they scatter light mostly toward the forward direction—that is, the path of the light beam is bent only slightly. Droplets and dust close to the line of sight between the viewer and the sun produce the impression of the dazzling white sky near the sun. Similarly, when the sun is behind a thin cloud, it is the light scattered in the forward direction by the droplets in the cloud that we see when we say the cloud appears white. A thicker cloud, heavy with large drops, will absorb much of the sun's rays and will appear gray or black, signaling the expected rain. The edges of a heavy cloud are thin, and light passing through them produces the cloud's "silver lining."

Sometimes heavy clouds are spread out over a great distance between us and the horizon. Then the atmosphere under the clouds is dark, and we see light only from the vicinity of the horizon, where the sun's rays are scattered from behind the clouds. Then the blue and green scattered light from that portion of the sky is largely lost over the long path toward us. The longer wavelengths in the red region are not present at all in the skylight, and so we see only light of intermediate wavelength, generally in the orange—an orange-colored sky low at the horizon beneath the gray clouds.

The scattering of light, creating the illusion of the sky, is the work of air molecules, each radiating short-wavelength light. The same molecules can be considered to act in concert to produce atmospheric refraction. Just as light travels more slowly in water or glass than in a vacuum, the speed of light is less (by about 0.03 percent) in air than in a vacuum. This causes the slight bending of starlight when it enters the atmosphere (see fig. 1.3) and the twinkling of starlight passing through turbulent air. On a hot day we can often see what appear to be waves at a distance in the air just above the ground. What we are seeing are the motions of hot air rising from the ground, made visible by varying curvatures in the pathways of transmitted light. A dramatic example of this phenomenon is the hot and rapidly moving air under a taxiing jet plane—often favored by directors of motion pictures.

A tar road or a concrete highway absorbs the sun's heat readily and may be much hotter than the atmosphere. The road creates a layer of hot rarefied air up to a foot or so above the surface, causing light rays to curve upward as they approach us. Thus we see objects above the surface as if they were inverted and located below (figs. 5.4 and 5.5). The illusion is of a puddle of water or a

Fig. 5.5. A mirage produced by a layer of warm air close to a hot road (Photograph by Robert Greenler)

Fig. 5.6. A complex mirage known as the *fata morgana,* photographed over the Arctic ice (Photograph by Robert Greenler)

pond reflecting sky or trees from above. This is the explanation of the most common type of mirage. Perhaps the Egyptians saw the Hebrews marching through the imaginary sea on the hot desert sands at the time of the Exodus. And did the Hebrews, looking back, see the Egyptians engulfed in the same mirage?

An opposite type of mirage occurs when there is a layer of cold air close to the surface—a situation found sometimes over a lake after a cold night. Then light rays curve downward as they approach the viewer. Objects on the ground may then appear to be looming in the air. Medieval "castles in the air" may have been seen in this way. Occasionally there is a more complex layering of air above the ground or a body of water—the temperature warm near the surface, colder a few feet above, and warm again above that. Then rays of light from a given point can curve both upward and downward as they approach the observer. The result is a view of vertical streaks across the landscape, often in motion (as air currents move) and fading in and out, leading imaginative viewers to see illusory castles, buildings, and cities (fig. 5.6).

Raindrops and Ice Crystals

The sun has moved westward in the sky and the west wind has blown dark clouds above us. Rain falls. If the sun is visible we can often see raindrops like sparkling crystals when we look close to the sun. What we are seeing are the sun's rays, refracted through small angles in the liquid of the drops, producing an image much brighter than the sky.

The rain clouds blow past us toward the east, and looking in that direction, we see one of nature's grand spectacles, the rainbow. Its arc is a portion of a circle, and the circle always has an angular radius of 42 degrees (fig. 5.7). If the sun is low, almost at the horizon, the arc we see is almost a full semicircle. If the sun is higher we see a smaller arc, and if the sun is more than 42 degrees above the horizon there is no rainbow (at least for viewers on the earth's surface). Observers on a hill may see parts of the rainbow below the horizontal, an arc larger than a semicircle; passengers in an airplane have sometimes seen a full circle, when there is rain in the air above and below them.

The colors fade smoothly from red on the outside to violet on the inside. If we are lucky we can see a secondary bow (fig. 5.8), fainter and larger (with an angular radius of 50 degrees) than the primary bow and with the order of the colors reversed. Still more rarely there are what are called *supernumerary arcs,* faint bands of pink, green, or violet located inside the main rainbow near the top of the arc.

The rainbow has been a source of wonder and mystery since the beginning of recorded time. It was a central symbol in the deluge myths of ancient Sumeria and Chaldea, as well as in the biblical story of the flood. Aristotle wrote extensively on the rainbow in his *De Meteorologica* and speculated that the rainbow is caused by the reflection of sunlight from the drops of water in a cloud.

The first scientific study of the rainbow—and other atmospheric phenomena—was provided by Theodoric, professor of theology at Freiburg in the fourteenth century. He studied in the laboratory the path of a ray of light in a spherical globe of water and proposed that the globe be thought of as a model for a single falling raindrop. The ray, entering the drop from the sun, is bent at the air-water boundary by refraction, then reflected from the inside surface of the drop; it then leaves the drop in a backward direction following a third contact with the surface. Thus, looking in a direction away from the sun, we see an enhancement of light due to these rays. Theodoric did not explain why this enhancement occurs only at a single angle from the sun's ray, or why the light is broken into colors.

It was René Descartes (fig. 5.9), the most influential French scholar of the seventeenth century, who, among his many writings in philosophy, mathematics, and science, provided the basic explanation of the rainbow, its colors, and its location in the sky. He showed mathematically that if one traces the path through a raindrop of parallel light rays entering the drop at different points on its surface, each will emerge in a different direction, but there will be a concentration of emerging rays at an angle of 42 degrees from the sun's rays—in agreement with the observed angular size

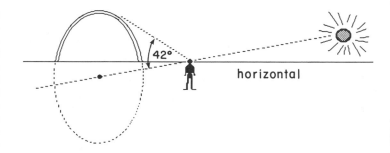

Fig. 5.7. The geometry of the rainbow. Rays reach the observer from a series of directions, all making an angle of 42 degrees with the line between the observer and the sun—more precisely, with the continuation of that line beyond the observer. The rainbow is not located at any particular set of points in space, but like an image in a mirror, it is an effect of directed light rays.

Fig. 5.8. A partial arc of a rainbow, showing both the primary bow and the secondary bow above it (Photograph by Robert Greenler)

Fig. 5.9. René Descartes (Courtesy AIP, Niels Bohr Library)

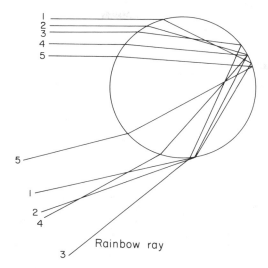

Rainbow ray

Fig. 5.10. Descartes's theory of the rainbow. The diagram shows parallel light rays entering a spherical raindrop. Only rays entering the upper half contribute to the rainbow effect. Let us follow the rays, one by one, from the top down to the middle of the drop, and consider what happens to rays that are reflected once after entering the drop and then reenter the air. We observe the following pattern: rays right near the top emerge in a direction almost reversed, a few degrees below the incident direction; rays a little farther below the top emerge in a more downward direction; finally the "rainbow ray" is one that emerges in a direction 42 degrees below the horizontal; rays entering the drop lower than the rainbow ray are now seen to emerge in a direction higher than 42 degrees. This variation, with 42 degrees being a minimum angle, leads to a bunching of rays at that angle, and hence an unusually high intensity of reflected light. The rainbow angle varies slightly with wavelength because the index of refraction of water does so, and therefore the colors appear at slightly different positions in the sky.

of the rainbow (fig. 5.10). Furthermore, since the refraction angle of each color will be slightly different—as Newton showed with his glass prism—the "rainbow angle" will be slightly different for each color, and the intense reflected rays will be spread apart into the solar spectrum.

Descartes was also able to show that light rays that are reflected twice inside the raindrop will come out concentrated at an angle of 50 degrees, corresponding exactly to the secondary rainbow sometimes seen outside the first (fig. 5.11). This would naturally be less intense than the primary, because a light ray always loses some of its intensity during each reflection or refraction.

More elusive than the rainbow is the *glory* (fig. 5.12), another effect of water droplets on light. Sometimes in mountainous regions a mist settles into a valley. The traveler, standing with the sun behind her, sees her shadow projected on the mist, her head surrounded by a wreath of light. Light rays enter each drop of mist and return in a direction almost reversed. The viewer sees the sun's rays returning to her eyes from close to the shadow of her own head. She cannot see the wreath about the head of anyone else. The glory is often seen from an airplane flying in sunlight above a cloud. The ring of light is seen below, sometimes colored, surrounding the shadow of the airplane on the cloud.

Descartes's theory of the rainbow depends on the assumption that the raindrop is a sphere; it does not depend on the size of the drop. The rainbow and the glory persist in the sky even as drops of different sizes move into the viewer's line of sight, provided the drops are spherical. Small drops are kept spherical by water's surface tension, which tends to force them into the shape

Fig. 5.11. The secondary rainbow is formed from rays that are reflected twice in the interior of a raindrop. The bunching in this case occurs for rays that enter the lower half of the drop and emerge 50 degrees below the sun's rays. Shown here are the rainbow rays for both the primary and the secondary rainbow.

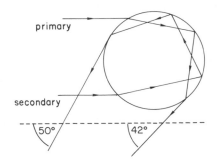

Fig. 5.12. The glory, a circle of light reflected from small water drops. This photograph shows the shadow of the airplane from which the observation was made. (Photograph by Robert Greenler)

Fig. 5.13. The 22-degree halo, with mock suns. The direct rays of the sun are blocked by the head of the person in the photograph. The halo is the faint circle of light centered on the sun, and the two mock suns appear in the sides of the halo. (Photograph by Robert Greenler)

of minimum surface area. Larger drops become flattened by the effect of air resistance. (None are teardrop shaped.)

Quite different is the behavior of water at lower temperatures when it becomes a solid. Then the molecules orient themselves in rigid geometrical patterns, which we know as crystals of snow and ice. At least eight different crystalline structures have been found for ice, but the most common one is hexagonal, six rectangular sides and two hexagons as upper and lower surfaces.

Sometimes on a cold spring day ice crystals form in high cirrus clouds, and sunlight shines through a shower of small needle-like hexagonal crystals. If you block the direct rays of the sun with your hand you may see a *halo,* a circle of light around the sun, with a radius of 22 degrees (fig. 5.13). The halo is caused by refraction of light, which enters one side face of the crystal and emerges from another side face. The ray emerges at different angles depending on how the crystal is oriented as it falls through the air. But, as in the case of the raindrop, there is a concentration of rays emerging at a particular angle—in this case 22 degrees—from the sun's direction, forming for us the image of the halo. Sometimes, if the crystals are relatively flat, they tend to be oriented not randomly but with the two hexagonal surfaces horizontal. Then, instead of light rays spread around the entire circle of the halo, they are concentrated at two points, approximately 22 degrees on the left and right sides of the sun—forming what are called mock suns, or *parhelia.* The halo and parhelia, being associated with refraction, may show color effects.

Snow crystals are even more varied than ice crystals. They are formed when water vapor freezes directly, without going

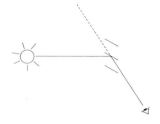

Fig. 5.14. Reflection of sunlight from flat ice crystals creates the sun pillar. Light rays reach the eye from a direction above the sun.

Fig. 5.15. A sun pillar (Photograph by Robert Greenler)

through the liquid stage—a process known as *sublimation*. Snow crystals in the form of flat disklike structures (or flat ice crystals), floating at an angle to the horizontal, are responsible for the *sun pillar,* a column of light rising vertically from a low sun (figs. 5.14 and 5.15). Light reflected from the bottom of a crystal strikes the eye from an angle higher than that of the direct sunlight, and so there appears a line of brightness above the sun.

Sunset

Afternoon draws on and the sun sinks toward the horizon. Now its rays travel through greater lengths of the atmosphere, and the effects of refraction and scattering become more pronounced. The sun appears more orange than yellow as more of the shorter-wavelength radiation is scattered by the atmosphere. Clouds in the west become gray or blue-gray as the weakened sunlight fails to penetrate them. They may be white on the edges, and at the bottom if the sun illuminates them from below. Thin clouds in the west will transmit much of the sunlight (really small-angle scattering by droplets in the clouds) and create dramatic patterns of red, orange, blue, and gray. Clouds in the east are still white or pink, reflecting sunlight that strikes them directly.

The sun sinks further and becomes blood red, its rays now so weakened that we can look at it directly. It is surrounded by a large diffuse semicircle of bright yellow-white sky, which fades into blue as the eye is raised. There is no sudden darkness as the sun falls below the horizon. For the sun's rays were already very weak, and most of the illumination of objects around us comes from the sky rather than directly from the sun (fig. 5.16). After sunset the sun continues to light the sky above us, its rays reaching the upper atmosphere from below our horizon. The day gradually fades into twilight. In the west the sky continues yellow or orange, due to scattering of sunlight through small angles by dust and water droplets in the air. These colors fade gradually into the deep blue and purple of the higher portions of the sky. Interestingly, the sky toward the east may be brighter than the sky at the zenith, because when we look up we look through a relatively short length of atmosphere and so the eye collects very little scattered light. Just above the eastern horizon the sky is darker; that is the direction of the earth's shadow, where the sunlight reaches no part of the atmosphere.

The most dramatic sunsets, with great splashes of crimson, orange, and purple in the western sky, occur when the air is most polluted: over industrial cities, where the air is filled with dust and smoke particles from burnt fuel, or after volcanic eruptions. In 1883 the island of Krakatoa in the Dutch East Indies erupted, sending millions of tons of volcanic ash and other debris into the

Fig. 5.16. Twilight. After the
sun sinks below the physical
horizon (at sunset), its scattered rays still reach us from
the atmosphere.

air. After three months fine particles drifted into the upper atmosphere, and brilliant red sunsets were seen in the Northern Hemisphere that year and for several succeeding years.

When the sun is low and its rays pass through the maximal length of the atmosphere, refraction of its light can produce surprising effects. Light from the lower edge of the sun is bent more than light from the upper edge. Rays from the lower edge then appear to come from a higher point in the sky, and the result is to make the sun look flatter, like an egg sitting upon the horizon.

If the air is calm, a temperature inversion can occur, a layer of warm air settling over a colder layer; or several layers of different temperatures can be maintained without mixing. Then the image of the sun can divide, mirage-like, into an upper and lower sector, which move together as the images approach the horizon. Because of refraction, we see the sun for a short time after it has gone below the physical horizon; its rays are bent around the earth's curvature. Further, as in the prism, rays of different colors bend to different degrees, red bending less, blue more—producing different colors at different portions of the low sun.

These color effects in the atmosphere are responsible for an intriguing, if transitory, phenomenon: the *green flash,* observations of which can be traced back to the ancient Egyptians. When sunset is observed at a low horizon (usually over the sea) in a clear sky, there is sometimes seen a flash of green light, just in the few seconds during which the sun disappears. Although normally it lasts just seconds, observation can be extended to a few minutes in polar regions where the sun slides parallel to the horizon during the transition from the long summer of daylight to darkness. Photographic studies have shown that a green tinge appears at the edges of the sun when it is about halfway below the horizon. This color spreads toward the center as the sun descends. Finally, like the smile of the Cheshire cat, all that is left is a splash of green on the upper edge of the sun, hovering briefly at or above the horizon. The green tinge on the sun is so small that it cannot be resolved by the eye (that is, the eye integrates this light with the yellow or red light from the rest of the sun and does not process a separate green color), and green is seen only in the last few seconds when the rest of the sun has disappeared.

Scientific study of the green flash was carried out in the nineteenth and early twentieth centuries. When the sun is very low it is not surprising that most of the blue is removed by scattering from the atmosphere, as explained by Rayleigh. Under some conditions, much of the yellow and orange is removed by absorption by molecules of water vapor, oxygen, and ozone. Thus only red and green remain. But because green light is refracted in the atmosphere more than red is, there can be an instant when the upper edge of the sun is below the actual horizon during which the red rays do not reach us; but the green rays are sufficiently bent to come around the earth's curvature, and they reach the observer in the few seconds after sunset. The red sun is below the horizon, and the green sun is above.

The Moon and the Stars

The sun dominates our visual life during the day, just as it dominates the motions of the earth, the air, and all of life. At night the moon is queen, spreading her light through the sky just as the sun does. The moon's light is so much fainter than the sun's that the moonlight scattered in the atmosphere is correspondingly fainter than the daytime sky. To the eye the sky looks black. But the night sky at the time of full moon is bright enough to obscure many stars that are visible at other times.

The moon's phases, as was understood as early as the fourth century B.C., occur because it is merely a reflector of sunlight. When the moon's position in the heavens is opposite to that of the sun, as viewed from earth, we see the full face of the moon illuminated. When the moon is close to the sun's direction, the illuminated face is away from us and we see either no moon at all or only a sliver, the narrow edge of the lit face. Thus we can see a full moon rise in the east just as the sun is setting in the west. We can see a half moon in the sky during the day, when it is at right angles to the sun. The crescent moon is in the west just after sunset or in the east just before sunrise. (There can also be a crescent moon just before sunset or just after sunrise, but in these cases it will not be visible because it is located very close to the brightly shining sun.) At night the moon is white or pale yellow, becoming deeper yellow, orange, or orange-red when it is low. As with the sun, this is caused by scattering of short wavelengths during the passage of light through long distances of atmosphere.

Sometimes at night a thick mist fills the sky and we perceive the moon shining through it. Then we can see a narrow ring of light, sometimes quite bright and tinged with color, around the moon. This is the *corona*, a fine example of diffraction in the atmosphere. The diameter of the corona varies from one to a few degrees and depends on the size of the water drops in the mist. Each water drop behaves like a circular pinhole, creating a diffraction pattern out of the incident moonbeams (just as a diffraction pattern is formed on film by the lens of a telescope). The image of the moon itself is a circle (even if the moon is not in full phase); it is the first region of constructive interference in the pattern; the corona is the second region of constructive interference. Larger and fainter circles outside the corona are sometimes visible, representing higher levels in the diffraction pattern. Color separation occurs because diffraction of the wave is different for different wavelengths. The colors and the corona itself are variable and unsteady because the size of drops in the mist changes over time.

At certain times of the month there is no moon at night, either because it is in the new phase or because it is below the horizon during all or part of the night. The sky is much darker, and many more stars are seen. At astronomical observatories all over the world the most desirable periods for viewing occur when the moon is not visible. Yet it is hard to see the night sky as it really is. In the populated areas of the world there are so many lights of cities, towns, and roads that the sky is filled with scattered light from man-made sources. Many people have never seen the moonless sky with its millions and millions of stars.

Airglow and Aurora

Still the stars do not fill the sky. The night sky appears black to us, each star a tiny white point of light. What are we looking at when we look at the black sky? Are we seeing through a thicket of stars to the edge of the universe? If we point a telescope in a direction between the visible stars, sensitive instruments pick up a faint glow across the sky, somewhat brighter near the horizon, sometimes unevenly distributed, as if from floating luminous clouds. It is called the *airglow,* light arising in the upper atmosphere, with a spectrum of bright lines. The intense yellow sodium lines have been identified in the airglow, as well as red and green lines of atomic oxygen[2] and weaker lines of calcium, potassium, and lithium. In addition the airglow spectrum contains wavelength *bands,* characteristic of excited molecules.

The spectrum of a molecule is more complex than that of an atom because the structure and the possible quantum states of a molecule are determined not only by the positions of the electrons around a nucleus but also by the arrangement of the several nuclei in the cluster. The major features of the spectrum are still determined by the possible electron states: there will be an electron ground state and various excited states. But for each electron state there is a series of vibrational states, corresponding to different degrees of vibration among the atoms in one molecule. An oxygen molecule, for example, can be thought of as two oxygen atoms held together by a spring, which permits vibrations along the line connecting the two atoms. As we would expect, this vibration is quantized, so that the possible vibrational energies form a series of discrete steps. The energy difference between vibrational levels is typically small compared with the spacing between one electron level and the next. A transition between one vibration level and another thus occurs by emission or absorption of a low-energy (infrared) photon.

In addition, a molecule in a gas is free to rotate, and so for each electron-vibration state there is a series of rotational states. Spacing between rotational energy levels is still smaller than that between electron and vibrational levels, so that the series of rotational levels, although quantized, is close to a continuous band. Correspondingly, the wavelengths of electromagnetic radiation emitted in transitions between electron-vibration levels are spread into bands. These molecular bands fingerprint molecules in the same way that spectral lines fingerprint atoms. Thus the airglow contains bands in the blue, violet, and ultraviolet, produced by excited molecular oxygen, and strong infrared bands produced by the hydroxyl radical, a combination of one oxygen atom with one hydrogen atom.

Research on the airglow has been used as a source of information about physical and chemical processes of the upper atmosphere. For example, although the amount of sodium in the atmosphere is minute, its yellow emission lines are so intense that they appear in the spectrum of the airglow. The presence of sodium in the air may represent a residue of salt crystals (a combination of sodium and chlorine) from the evaporating spray of the oceans; or salt may enter the atmosphere from a shower of volcanic dust. Sodium, calcium, and other elements may also reach the upper atmosphere from outer space—from interplanetary dust or from the larger chunks of solid material that enter the atmosphere to produce meteors.

Although airglow was first detected at night, it has been verified that similar emissions occur during the day, in fact more intensely than at night.[3] The source of excitation of gas atoms and molecules must ultimately be sunlight. During the day free atoms and molecules in the atmosphere absorb ultraviolet light from the sun, thereby being raised to excited energy states. They subsequently lose their excess energy by emitting characteristic radiation (sometimes of longer wavelength), spectral lines for atoms and bands for molecules. Most of the sun's ultraviolet does not reach us on the surface of the earth because it is absorbed by the ozone layer. But at heights above 60 miles the air is exposed to the full strength of this radiation and this directly drives the day airglow. A number of other processes are involved in day airglow, including ionization (the removal of one or more electrons from an atom) and molecular dissociation (the separating of a molecule into atoms), which are both caused by ultraviolet radiation.

It is more difficult to explain the occurrence of night airglow, when no solar radiation enters the atmosphere. It is probably associated with photochemical processes—chemical reactions in which one of the products of the reaction is in an excited state and subsequently de-excites by emitting light. During the day, for example, the sun's ultraviolet produces atomic oxygen by breaking up molecular oxygen into its two separate atoms. These oxygen atoms produce chemical reactions when they collide with other constituents of the upper atmosphere. But because the atmosphere is so rarefied, collisions are infrequent, and the reactions proceed slowly, lasting until night. The products of these reactions are excited atoms and molecules which decay, emitting the characteristic light of the night airglow.

The airglow has been called a permanent aurora, and indeed the spectra seen in auroras (the *aurora borealis,* or northern lights, and the *aurora australis,* or southern lights) are similar to that of the airglow. There are green and red lines of atomic oxygen, bands in the blue and violet associated with ionized molecular nitrogen, and occasionally lines of atomic hydrogen.

The aurora (fig. 5.17), normally visible only in latitudes close to the poles, provides some of the most dramatic demonstra-

Fig. 5.17. The northern aurora (Courtesy National Oceanic and Atmospheric Administration)

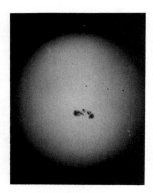

Fig. 5.18. Sunspots (Courtesy Yerkes Observatory)

tions of nature's splendor, in light, color, and motion. There are broad arcs stretching across the sky from east to west, more diffuse glows over almost the entire polar sky, and more sharply defined vertical rays; these patterns sometimes twist themselves into loops and folds, resembling hanging curtains, and into more irregular shapes. Arcs may alternately dim and brighten for a period, or they may gradually brighten and move upward, finally developing rays and loops as the night lengthens. There sometimes appears a rising wave of brightness (known as a "flaming aurora"). White, yellow-green, and red are the most frequently seen colors; blue, violet, and yellow are rarer. At the peak of an active display the patterns may vary so rapidly that they cannot be encompassed by the eye; at times, in the hours before dawn, for example, the aurora may be simply a diffuse light across the sky, slowly fading as morning approaches.

Scientists are quite far from being able to explain the patterns of the aurora and their fluctuations through the night. But the basic mechanism of light production has been understood and ties in with numerous other areas of research, including the chemistry of the upper atmosphere and the physics of the surface of the sun. The intensity distribution of auroras over the earth correlates most closely with the geomagnetic poles rather than with the geographic poles. Maximum intensity of the aurora borealis is in a region around 23 degrees south of the north geomagnetic pole (including Alaska and northern parts of Canada, Scandinavia, and Russia). Auroras are particularly strong when the earth's magnetic field is highly disturbed—that is, when it varies rapidly—and they are frequently correlated with a high degree of ionization in the ionosphere. These effects are also associated with unusual activity of the sun, including sunspots (fig. 5.18) and solar flares. Rapid agitation of gases around sunspots is sometimes followed by violent ejection of flaming gases above the sun (at temperatures of millions of degrees), accompanied by a flow of electrically charged particles into interplanetary space. It is now known that there is a continuous flow of charged particles, protons and electrons, from the sun into space. This flow, known as the *solar wind,* travels in all directions in the solar system (and has been studied by American spacecraft not only near the earth but also near other planets). Those particles that approach the earth are trapped by its magnetic field and spiral in along magnetic lines of force. The flow of charged particles is therefore concentrated at the magnetic poles, and it is closely tied to the degree of activity on the sun. Active auroras correlate with the eleven-year cycle of sunspot activity. (Sunspots are regions of the sun's surface that are cooler—hence less bright—than the surrounding surface, and so sunspot activity is associated with an increase in the upwelling and swirling motions characteristic of the surface.)

Thus it is natural to expect that charged particles entering the upper atmosphere from space are responsible for the lights of the aurora. In recent experiments, rockets have been flown directly into the illuminated portions of the sky (between 60 and 600 miles above the earth's surface) to measure the flow of charged particles. It has been found that certain patterns of auroral light correlate well with the intensity of electron streams through the atmosphere. Thus it is clear that the light of the aurora is produced by charged particles in much the same ways as the airglow is produced by the sun's ultraviolet. That is, highly energetic electrons from the sun collide with atoms and molecules, exciting them to various energy levels, after which the atoms fall back to lower levels, emitting light in the process.

Study of the aurora exemplifies the highly interdisciplinary nature of modern research. The study of the sun with its swirling gases, intense heat, and magnetic fields, the solar wind, the magnetic field of the earth generated in its molten core, the composition of the upper atmosphere, and the interactions of particles at the molecular and atomic level—all these disciplines are interwoven in the study of the northern lights. Paradoxically, as scientists become more specialized, the sciences become more interdependent and unified.

There is still more to be seen in the night sky. A faint illumination, concentrated along the Milky Way, arises from integrated starlight. It is the totality of light from all the stars that are too faint to be seen individually. It has a continuous spectrum, like that of the sun. A second source from outer space, visible to the eye on very dark nights, is the *zodiacal light,* so-called because it is most intense around the arc of the zodiac. (This is the great circle across the sky in which the twelve constellations of the zodiac lie.) The arc marks out the plane of the earth's orbit; close to this plane are found the orbits of the other planets and also a vast disklike cloud of dust, debris left from the formation of the solar system. Sunlight reflected from this interplanetary dust produces the zodiacal light: it is a faint pyramidal form seen above the western horizon after sunset and above the eastern horizon before dawn.

But we have allowed ourselves to be carried beyond the atmosphere, and indeed beyond the scope of this chapter. It is midnight and the last trace of the western zodiacal light has sunk below the horizon. We look toward the zenith where airglow is faintest, and we are transported from the bounds of our planet. From sky and earth nothing reaches us, except the message of the stars.

The Stars

6

To Christopher Columbus the stars were a guide across the Atlantic. To the astronaut the stars are a guide across the solar system. To the city dweller the stars are barely visible. Yet the stars might have been everything. On the grand scale all the light of the universe is of the stars, and until recently we might have said all matter and energy lies in the stars.

The stars are also everywhere. They are so numerous and appear to be so homogeneously spread out in space that the nineteenth-century astronomer Wilhelm Olbers argued that the night sky ought to be completely filled with light—indeed that every point in the sky should be as bright as the sun. The discrepancy between this prediction and obvious fact became known as *Olbers' paradox*. We will see in the next chapter how it is analyzed by modern astrophysics.

The stars are not only everywhere; they are also immovable. True, we appear to see stars move through the sky during the night just as the sun does during the day; some stars rise in the east and some set in the west. But if we choose two stars separated in the sky by a given angle and observe them for many years, we find that that angular separation never changes. The same constellations—patterns of stars—that were named by the ancient Greeks appear to us, thousands of years later, unchanged.[1] Thus we expect that it is not the stars that are in motion but the earth and that the nightly motions of the stars are just a reflection of the daily rotation of the earth on its axis.

The Distance to the Stars

Is a star anything more than a point of light? To the naked eye the star has no shape, no dimension, no color, not even a distance that we can fathom. Yet with the tools and equations of modern science we can extract the message of starlight and deduce the presence of billions of fiery suns much like our own.

Until we learn the distance to a star, we know almost nothing about it. We have seen that optical instruments measure only the angular size of objects and that only indirect clues tell us whether

119

Fig. 6.1. Parallax, the apparent motion of a near star, *S,* due to the motion of the earth in its orbit. During the summer *S* appears between distant stars *F* and *G.* During the winter *S* appears between *A* and *B.* To view the star a telescope on earth must be aimed in a different direction in summer and winter. The angle between these two directions, α, is called the *parallax angle.* Typical values for α are very small, smaller than 1 *second of arc* (1/3,600 of a degree). The definition of the *parsec* is that distance for which the parallax angle is exactly 1 second.

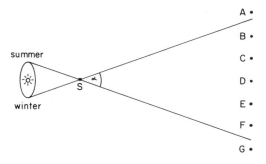

an object is small and nearby or large and distant. How much more mysterious is a single point of light in the dark sky! Is it a diamond at the rooftop or a glowing furnace billions of miles away?

Definitive answers to this question were found only in 1838, when telescopes became powerful enough to observe the *parallax* of nearer stars. Parallax is an everyday concept, and it is the key to calculating the distance to the stars. Hold your forefinger at arm's length and view it with only your left eye open, facing a window across the room. Now hold the finger still but view it with your right eye only. The finger has apparently moved, perhaps one-third of the way across the window. Let your finger represent one of the stars nearest to our solar system, and let the window represent several stars much farther away. If we could view the nearer star from two different vantage points—like your two eyes—we would detect its parallax, or apparent relative motion. In fact we *can* so view the stars, by observing at two different points along the earth's orbit around the sun (fig. 6.1). Now take the analogy one step further. Observe again the parallax of your forefinger in front of the window. But this time hold your finger closer to your eyes, let us say 12 inches away. You will see an even greater parallactic motion; the finger seems to move the entire width of the window. Thus the amount of parallax depends on how far your finger is from your nose. In the case of a star also, the amount of parallax depends on how far the star is from the earth. Astronomers were first able to measure the distance to near stars by observing their parallax with respect to much more distant stars.

They found the nearest stars to be several hundred thousand times as far from us as the sun, distances of the order of tens of trillions of miles. For convenience in discussing stellar distances, astronomers have introduced a unit called the *parsec,* approximately equal to 19 trillion miles.[2] The nearest star, α Centauri, is 1.3 parsecs away. Because these distances are so large the parallax of stars is extremely small and can be detected only with powerful telescopes. The motion scaled down to the forefinger at arm's length would correspond to a few ten-thousandths of an inch along the window.

The phenomenon of parallax was understood before the nineteenth century, but parallax of stars had not been observed. Thus scientists had argued in reverse: if parallax is not seen, then it must be that the earth does not move. The lack of observation of stellar parallax was a telling argument during the great historical debate between proponents of the heliocentric and the geocentric theories of the solar system. This debate reached its height in the period from 1543, when Copernicus published *De revolutionibus* (his theory of the motion of the earth and the planets around the sun), to 1642, with the death of Galileo. Although Galileo observed the stars with his telescope and cogently supported the heliocentric theory, opponents responded that the earth could not move about the sun, for if it did parallax of near stars would be seen. By the nineteenth century other lines of evidence had substantiated the theory of Copernicus, but the observations of stellar parallax, almost 300 years after his death, was the first direct confirmation of the fact that the earth moves.

What do we learn from this datum: that a star is trillions of miles away, let us say about 10^5 times as far as the sun? Suppose that same star were moved to a distance of 93×10^6 miles, the actual distance to our sun. We know how to calculate how bright the star would appear at this distance, because light intensity falls off as the square of the distance. If a star is moved 10^5 times closer to us, it appears 10^{10} (the square of 10^5) times as bright. Knowing the observed brightness of a star—say, the amount of light energy per second that strikes the lens of a telescope—we can thus calculate how bright it would be at the sun's distance. We find that the brightness of a typical star at 93×10^6 miles is about the same as that of the sun. On this basis we may deduce that the sun is really a star—the same kind of celestial body as the billions of stars in the sky. Its only distinction is that it happens to be the one around which we orbit, and it is therefore very close.

Further, we are led to the new concept of the *intrinsic brightness* of a star, which we can think of as the brightness a star *would* have at 93×10^6 miles. It is a measure of the amount of light leaving the star's surface per second; it reports something about the star's inner working.

The Classification of Stars

Is it fair to think of a "typical" star, as we have done above? Although stars vary in intrinsic brightness by factors up to a few thousand, there seem to be limitations to this range. There are no stars as bright as a galaxy; there are no stars as bright as a light

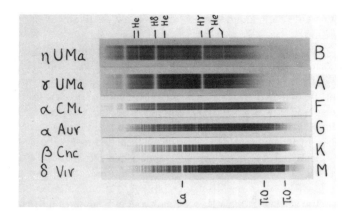

Fig. 6.2. Spectra of stars. These photographs are negatives, so that dark lines appear as bright lines on the film. Spectra of six different stars are shown, ordered according to temperature, the hottest at the top. In general cooler stars have more spectral lines. This is because at high temperatures the random motions of atoms are more energetic, and in collisions many of the outer electrons are removed from the atoms. Thus the lines associated with transitions normally made by those electrons are not present. (Courtesy Yerkes Observatory)

bulb. To some extent we will see that many stars can be called typical, or "normal."

The same is true for color. To the naked eye most stars are white, although on a clear night we might detect a reddish or bluish tinge to some stars. The eye does not easily detect the color of a star, because we observe at night with the rod system. The spectroscope, however, allows us to determine the predominant colors in starlight and the relative amounts of light of all different wavelengths. Like the sun, the stars have a light intensity curve centered around one predominant wavelength, with lower intensity at larger and smaller wavelengths. The peak wavelength (with certain very rare exceptions) is always in or close to the visible region—red, green, yellow, blue. Given the whole electromagnetic spectrum, from long-wave radio to x-rays and gamma rays, stars radiate in a narrow band around that which is visible to us.

It is primarily intrinsic brightness and color that lead astrophysicists to important generalizations about stars. A star is very close to a perfect blackbody,[3] and therefore its color correlates directly with its temperature. Blue stars have temperatures in the region of 20,000° C, red stars around 3,000° C. The sun, an intermediate star, has a temperature of 5,800° C. These high temperatures, hot enough to vaporize any of the elements, represent only the temperatures of the outer boundaries of the stars, from which light comes into space. The temperatures deep inside the stars are not known, although theory indicates they are not in thousands, but in millions of degrees.

A star's spectrum also shows dark spectral lines, like the sun's (fig. 6.2). The wavelengths of these lines are found to be the same as those first found by Fraunhofer in the sun: lines of hydrogen, helium, iron, calcium, and so on. Not all lines in stellar spectra have been identified, but we have every reason to believe that the matter of the stars consists only of the elements found on the earth and the sun. The stars are primarily hydrogen and helium, with relatively small quantities of other elements.

The first approach to classification of the stars is based on an analysis developed independently by Ejnar Hertzsprung (fig. 6.3) in Denmark and Henry Norris Russell in the United States. They drew a diagram with the star's surface temperature along the horizontal axis (higher temperature toward the left) and intrinsic brightness along the vertical axis (fig. 6.4).[4] Then each star was recorded as a point on this graph. The correlation they observed was that the brighter stars were at higher temperatures, radiating at shorter wavelengths (toward the blue), and the dimmer stars at lower temperatures, radiating at longer wavelengths (toward the red).

Russell's diagram showed stars in the neighborhood of the solar system. About 90 percent of these stars fall close to a single smooth curve in the plane. These are said to be on the *main sequence*. It was thought originally that this represented a sequence of changes that a star went through during the course of its lifetime. We have since gathered evidence that each point on the main sequence represents, not a stage in a star's life, but rather the point (characteristic color and brightness) at which a star spends almost all of its life; each different point on the main sequence corresponds to stars of a given mass.

A star forms when a huge cloud of gas (mostly hydrogen) begins to collapse and coalesce under the influence of the attractive force of gravity. As it collapses it becomes denser and hotter, until eventually it becomes sufficiently hot at the center for nuclear burning to begin. Thus the star is born. For a more massive original cloud, the gravitational force is greater, more heat is re-

Fig. 6.3. Ejnar Hertzsprung (*right*) and Harlow Shapley (Courtesy AIP, Niels Bohr Library)

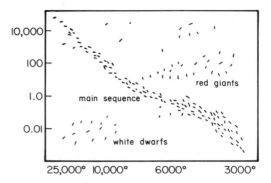

Fig. 6.4. The Hertzsprung-Russell diagram. The vertical scale is the star's luminosity relative to that of the sun. The horizontal scale is the *absolute temperature* of the star's surface, temperature increasing to the left. Absolute temperature is equal to the Celsius temperature plus about 273 degrees.

leased in the collapse, and the resulting star is hotter. The more massive and hotter star radiates predominantly toward the blue. It is larger and so radiates more energy; it is brighter. Hence the correlation observed in the Hertzsprung-Russell diagram. If the mass of the original cloud is too small, it may collapse, but it will not become hot enough to initiate nuclear burning. It will never shine. Any mass less than about one-hundredth of that of the sun will fail to become a star. Jupiter, for example, the largest planet in our solar system, is about one-thousandth of the sun's mass. It missed by a factor of ten.

Thus there is a lower limit on the mass of stars, and it follows that there is a lower limit on brightness, a lower limit on temperature, and an upper limit on the wavelength of emitted light. We do not yet have a complete explanation for the upper limit on the size—and brightness and temperature—of stars. We do know that what seems to happen to a very large mass is that, instead of forming a single star, it breaks up and forms two stars, *binaries,* which orbit around each other. Binaries are very common in the heavens, and frequently we see three or more stars orbiting mutually. There are also large clusters of up to tens of thousands of stars.

Nuclear Burning

Each star begins its life at some point on the Hertzsprung-Russell diagram, with a brightness, temperature, and color determined by its mass. It then remains unchanging and incredibly stable throughout the billions of years of its existence, slowly and evenly burning hydrogen fuel in its central core. The source of this stability is also the source of the energy that powers the sun and other stars, and sends photons through space. It is nuclear energy, or, as it is loosely called, *nuclear burning.*

Nuclear burning is a phenomenon entirely different from conventional chemical burning, as of a fuel like coal or wood.

124

Just as the explication of chemical burning in the 1770s marked the beginning of modern chemistry, the study of nuclear burning in many ways marks the beginning of the modern age. The association of nuclear energy with stars had to await two events: Einstein's discovery in 1905 of the famous relation between mass and energy and Rutherford's discovery in 1911 of the atomic nucleus. In chemical burning the electrons of the atoms involved in combustion are rearranged. For example, when wood burns, various compounds containing carbon interact with oxygen from the atmosphere and produce (largely) carbon dioxide. The nuclei of the carbon and oxygen atoms are unaffected. Indeed, until the twentieth century scientists believed that the atom was an indestructible unit of matter—that carbon atoms could combine with other atoms to form molecules, but they would never cease to be carbon.

The earliest scientific explanation of the sun's energy was that it was produced by chemical burning; that is, the sun is a giant ball of fuel, burning continuously. Knowing the mass of the sun and the rate at which it is producing energy, we are able to calculate its life span—how long the fuel will last. The result is about 3,000 years.

A more likely candidate for the source of the sun's energy is the energy associated with its self-gravitation. We usually think of gravity as the force exerted by a large object on a small one. The earth's gravity pulls down a tennis ball. But every bit of matter exerts a gravitational attraction on every other bit of matter, and in a massive body like the sun these forces add up to a huge overall inward force. In the nineteenth century Hermann von Helmholtz and William Thomson, who later became Lord Kelvin, proposed that the sun, by gradually shrinking, could slowly convert its gravitational energy into radiated light and heat. On this basis the sun's lifetime could be as much as one hundred million years. The ability to make a quantitative prediction using this theory led to its eventual failure, for early in the twentieth century fossils were found that could be dated to more than one hundred million years in the past. Indeed we now have evidence of microorganisms existing more than three billion years ago. So the earth and the sun, which supports life, are at least that old. The sun cannot be powered by chemical or gravitational energy.

In contrast to chemical burning, nuclear processes involve rearrangements of the protons and neutrons in nuclei to form new nuclei. Atoms can indeed be created or destroyed or changed into other atoms. For example, a carbon nucleus containing six protons and six neutrons could gather a proton and a neutron from a second nucleus with which it collides and become a nitrogen nucleus. The ancient quest of the alchemist to turn "base" metals such as lead into gold can never be accomplished by chemical means; but lead can be (and in the last few years has been) turned

125

into gold and other elements in nuclear research laboratories. (Needless to say, these are not profit-making operations.)

Nuclear reactions of many kinds were studied in detail, beginning in 1911 with Rutherford and continuing during the 1920s and 1930s with the work of James Chadwick, C. T. R. Wilson, F. W. Aston, and others in England, Irène Joliot-Curie and Frédéric Joliot (daughter and son-in-law of Madame Curie) in France, and Enrico Fermi and his colleagues in Italy, among others. In this work it soon became clear that large amounts of energy could be created or destroyed, quantities of energy that are typically a million times greater for a single atom than are produced in chemical reactions. When nuclei rearrange themselves after a collision, the mass of the products is not the same as the mass before the collision. If mass is lost, it appears in the form of newly created energy. In Einstein's equation, $E = mc^2$, m stands for the amount of mass loss, E for the amount of energy gained, and c for the speed of light. The equivalence of mass and energy, deduced on purely theoretical grounds in Einstein's 1905 paper on relativity, receives its most substantial confirmation in nuclear reaction experiments.

By the 1920s the British astronomer Arthur Eddington (see fig. 10.5) was advancing the idea that the sun's energy could be explained as energy released during nuclear collisions. In 1938 nuclear physicist Hans Bethe (fig. 6.5) proposed a detailed theory of nuclear burning in the hot central core of the sun. From the dark lines in the spectrum we know that the sun's outer region is made up primarily of hydrogen, with a large admixture of helium. Bethe proposed that the sun generates energy through the collisions of hydrogen nuclei, which in a series of steps combine so that four hydrogen nuclei form a single helium nucleus. (In the course of these reactions two of the protons are transmuted into neutrons; thus the helium nucleus—two protons and two neutrons—is assembled. The positive electric charge of the protons is carried off by positrons, which are the antiparticles of electrons.) This coming together of light nuclei to form a heavier nucleus is called *nuclear fusion*. Scientists have imitated it in the laboratory, and it may become a practical source of energy.

The helium nucleus is known to have a mass slightly smaller than that of four hydrogen nuclei, and this missing mass is turned into energy. Energy appears in the form of the random motion of the products and in the form of radiation—gamma rays created in the center of the sun.

It is only in the central core of the sun, or a star, where the temperature is many millions of degrees, that these nuclear processes take place. Energy works its way gradually from the core toward the outer edge by a process of absorption and reemission of radiation by the atoms of the star's gaseous matter. Short-wavelength gamma rays are absorbed by atoms, which then emit radiation of longer wavelength, and this process is repeated many times as the radiation moves outward. (Energy is also transported by convection, the outward flow of hot gases.) Finally, when radiation leaves the outer surface of the star, its wavelength is in the region of visible light—determined by the relatively cool temperature of the star's exterior.

Fusion of hydrogen in the center of the star not only produces energy but also creates a tremendous gas pressure in the star's core—like the hot wind that surges from a raging fire. This pressure exactly counterbalances the force of gravity, which pulls the matter of the star in toward the center, and so the star is kept in a state of dynamic equilibrium. The star is constant: constant in its energy output, in its size, and in its temperature. Equilibrium is established in the early stages of the star's formation. As the gas cloud collapses under the influence of gravity, the center becomes hot enough for nuclear fusion to begin, and the rate of this fusion increases until just that point at which the outward pressure equals the inward force of gravity; the collapse of the star then ceases. In other words, equilibrium here does not exist by chance. It exists because when a system is out of equilibrium, it must change; and it continues to change until equilibrium is reached, at which point there is no further change. (It is like lying down on a waterbed. Your body sinks into the mattress until the upward pressure of the mattress exactly balances your weight. If the pressure did not balance your weight at a given indentation, you would simply sink in a little further until you reached the point where it did balance.) When we survey the physical world it is not surprising that much of what we see is in the state of equilibrium.

How do we know that this theory of stellar energy is correct? Certainly nuclear burning is the only source of energy we know of capable of generating the prodigious power that emerges from the sun and the stars. We can also demonstrate in the laboratory the nuclear reactions that are supposed to occur in the star's core, verifying that they indeed occur and determining the rates at which they proceed. But beyond that we are left with the realization that we can never actually see into the center of a star, or even into our own sun. The sun is opaque; light produced in its interior never reaches into space or to us on earth. We see light only from the outer surface.

There is one source of information about the central core of the sun that does reach us on earth, and scientists have attempted a difficult and extended experiment to exploit this source. The carrier of information is the *neutrino*, one of the strangest discoveries of twentieth-century physics. The neutrino is a particle, not normally a constituent of matter as we know it but produced in some nuclear processes. When it is produced it bears the distinction of being the only known object, other than the photon itself,

Fig. 6.5. Hans Bethe (Photograph by Barrett Gallagher, Cornell University; courtesy AIP, Niels Bohr Library)

that travels at the speed of light.[5] But unlike light, the neutrino interacts extremely weakly with other kinds of matter. Thus, while a beam of light will never penetrate a brick wall and only a small part of it will penetrate a piece of paper, a beam of neutrinos (like x-rays) *will* penetrate matter readily. A large fraction of a beam of neutrinos will penetrate the entire earth and the entire sun. Neutrinos are created as by-products in the nuclear reactions at the center of the sun. Most of them will travel through the body of the sun and out into space. By studying these solar neutrinos we can indirectly "look" at the process of energy generation in the core of the sun. In 1965 an experiment was set up by Raymond Davis of Brookhaven National Laboratory to try to detect these elusive particles, which, just because they interact so weakly with matter, are very difficult to capture. A huge vat of cleaning fluid (used because chlorine, one of the constituents of cleaning fluid, is a particularly efficient nucleus for catching neutrinos) was set up in an abandoned gold mine in South Dakota. The object was simply to see if the number of neutrinos observed agrees with the number predicted by the theory of solar burning. The results of this experiment are largely disappointing. The number of neutrinos seen is considerably smaller than expected according to theory. Despite this result, most scientists believe that the basic concept of nuclear production of the sun's energy is correct and that some modified version of the existing theory will eventually be found that is consistent with the results of the Davis experiment. Improved versions of the experiment are being explored.

A Star's Death

The sun's equilibrium was established four to five billion years ago, and it has not changed since then. It will last until the hydrogen fuel at the sun's center is used up. The total lifetime of the sun is calculated to be about ten billion years. We have five to six billion years to go until our star begins to undergo various drastic changes associated with its impending death.

Since astronomical time scales are so large relative to a human lifetime—or even relative to the years of recorded history—it is difficult to reconstruct any of the pivotal events in the history of the sun or the stars. We see the sun and the stars at equilibrium and unchanging. Nevertheless, by studying enough stars we can observe some of the steps in the death of a star, when its hydrogen fuel is exhausted; we can also see evidence of stars being born. When the Hertzsprung-Russell diagram (see fig. 6.4) is drawn and the near stars are placed on it according to color and brightness, not all the stars fall on the main sequence. There are exceptions; about 10 percent of the stars do not follow the correlation of the others. Most of these exceptions are stars that have low luminosity and at the same time have colors toward the short-wavelength end of the spectrum. Lacking the red color that dim stars usually possess, they are called *white dwarfs*. An even smaller number of stars are found in the opposite corner of the diagram: these are red stars that are unusually bright and are called *red giants*. Through a combination of observation and theory astronomers have concluded that red giants and white dwarfs are, in that order, successive later stages in the evolution of stars.

After the hydrogen fuel is burned, there is no longer any outward pressure from the star's center to balance the inward pull of gravity. The star begins to collapse again—as the gas cloud did on its way to becoming a star—and the core becomes still hotter. At the same time, the outer region of the star begins to expand and becomes cooler and more rarefied. Eventually the outer envelope of the star reaches a diameter of about one hundred million miles (compared with a typical diameter of one million miles for main-sequence stars like our sun). When this happens to the sun, it will extend out to the orbits of the inner planets. And though the temperature will be something like a relatively cool 3,000° C, still it will be hot enough to extinguish life on earth. Stars in this stage of evolution are the red giants. It is a temporary and unstable stage, lasting a relatively short time; hence we do not see many red giants.

While the outer matter of the star is expanding, the collapse of the core eventually raises the central temperature high enough to make possible a new kind of nuclear fusion: the fusion of three helium nuclei to form a single carbon nucleus. This new nuclear burning reestablishes equilibrium in the central core, and the star remains in a stable condition as long as helium fuel is available. But the outer regions of the red giant become cooler and cooler, and eventually mass begins to detach itself from the star and float out into space. Sometimes this mass loss is in the form of a huge shell of matter (representing perhaps 20 percent of the star's original mass), which separates from the core and expands into space. Such clouds of gas have been observed (fig. 6.6) and are called *planetary nebulae*. (A misnomer: in early observations they looked like the outer planets; "nebula" is the Latin word for "cloud.")

But when the outer layers of the star are gone, only the core is left. The core is small, something like twenty thousand miles in diameter, and very hot. This is what we see as a white dwarf. In the core of the star helium burning is taking place, and eventually the helium fuel is completely consumed. The white dwarf is then left in space, white-hot, with no inner source of energy. As its energy continually streams out into space, the star becomes cooler and cooler, dimmer and dimmer, and the wavelengths of its radiation become longer and longer. Finally it will become so cool that its radiation can no longer be detected and it will be what we call a *black dwarf* (not to be confused with a black hole). Space may in time be strewn with such burned-out remains of stars. They emit no light and so we would not see them. Astronomers believe, however, that our galaxy is not old enough for many of its stars to have yet reached the black dwarf stage.

The black dwarf and the white dwarf are strange objects in another sense. These stars are typically about the size of the earth (earth's diameter is about eight thousand miles) and other planets.

Fig. 6.6. Planetary nebulae. (*top*) A ring-shaped nebula in the constellation Aquarius (Courtesy Palomar Observatory); (*bottom*) the Dumbbell nebula in Vulpecula (Courtesy National Optical Astronomy Observatories).

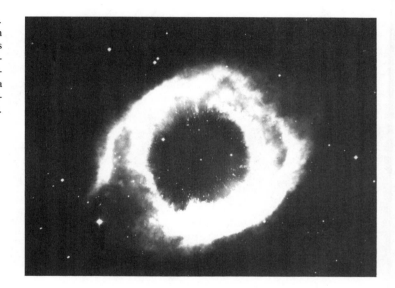

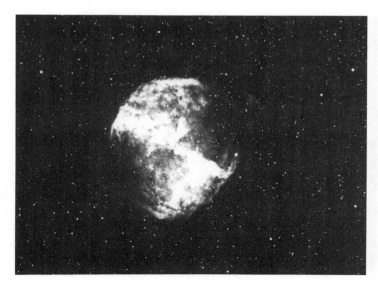

Like the planets, they have no internal source of energy.[6] But the white dwarf is the result of the gravitational collapse of a substantial fraction of the matter of a star. Thus a huge amount of matter, much more than the mass of the earth, has been compressed into a volume similar to that of the earth. The white dwarf is therefore extremely dense, about one hundred thousand times as dense as the planets. The materials we are familiar with on earth—water, stone, iron—are all of different densities; some, like lead, are considerably denser than others. But the densest element on earth, osmium, is only 22.5 times the density of water. The white dwarf is therefore an entirely different kind of matter, made indeed of protons, neutrons, and electrons, but in an entirely different form, unlike the solid, liquid, and gas that we are familiar with on earth. If you stood on such a star its force of gravity, because of the great density under your feet, would be a hundred thousand times greater than it is on earth. Indeed you would not stand at all, but you would be flattened to a puddle.

Even stranger than the white dwarf in the annals of stellar astronomy, and playing a similar role in the story of the evolution of stars, is the *pulsar,* discovered in 1967 in the radio portion of the electromagnetic spectrum. The pulsar's radio signal takes the form of rapid pulses: a burst of radiation for an interval of about a hundredth of a second, then quiet for about one second, another burst for a hundredth of a second, another quiet period, and so on. The transmission comes from a fixed direction in space and continues in this on-off fashion for months of observation, the timing precise and unvarying. What is so astonishing is not just the pulsation but the time scale. Stars normally exhibit a stability and constancy over billions or hundreds of millions of years. Some vary in brightness over periods of a few days or hours. But normal astronomical objects, even white dwarfs, are simply too large and massive to undergo changes in periods of seconds.

The most satisfactory early theory proposed that a pulsar was an object radiating in only one direction and rotating on its axis.[7] The signal from such an object would be viewed on earth as a pulsation, just as a rotating searchlight is seen by observers only when the beam of light crosses their eyes. But how could an astronomical object rotate on its axis once every second? Theorists estimated that a pulsar could pulsate at such a rapid rate only if it were extremely small—about 10 miles in diameter. It is not difficult for us to imagine astronomical objects of this size—some of the solar system's asteroids are only a few miles across—but the pulsar is an object we can detect from trillions of miles away and so is expected to be something like a star in total mass. Now if the matter of a typical star is compressed into a sphere of diameter 10 miles, the result will be an object whose density has the incredible value of 10^{15} times the density of water. We are taken another step

Fig. 6.7. The Crab pulsar. Two photographs of the same region of the sky are shown: (*left*) when the pulsar is not emitting light and (*right*) when the pulsar is emitting light. The other bright objects are nearby stars. (Courtesy Lick Observatory)

down the road beyond the white dwarf, which we saw as a new kind of matter, one hundred thousand (10^5) times denser than typical solids and liquids on earth.

Soon many other pulsars were discovered, not all pulsing with a complete on-off cycle, but varying regularly with periods ranging from a few seconds down to about one-thirtieth of a second. Some pulse in visible light (fig. 6.7). It was even possible to detect the slowing down of some pulsars; their periods become slightly longer after a few years. This can be taken as confirmation of the theory that they are rotating objects, since, as the pulsar radiates into space, it must lose energy, and this loss of energy causes its rate of rotation to decrease.

Bizarre though the pulsar is, it was not long before it found a place in the history of stars. For the final stages in a star's life depend crucially on its mass, and it is only for relatively light stars (up to about twice the sun's mass) that the final stage is the white dwarf, as discussed earlier. For more massive stars the story is more complicated and more spectacular. Driven by more powerful gravitational attraction, a heavier star becomes hotter at the core and is able to initiate nuclear burning of elements heavier than helium. Nuclear fusion proceeds, creating heavier and heavier nuclei at higher and higher temperatures, until finally the star reaches the limit of its energy-generating capacity,[8] and it collapses. This is a cataclysmic event, nuclei hurling themselves toward the center, colliding, and recoiling outward in a violent explosion. In some way not completely understood, a portion of this matter coheres in the center and forms an extremely dense core, which we later see as a pulsar. Although we do not have a complete theory of how light and radio are emitted from pulsars, we do understand its rapid rotation as a natural consequence of rotation of the original star. The pulsar is believed to be a state of matter similar to the interior of an atomic nucleus but containing only neutrons. It is called a *neutron star.*

A massive star in the act of explosion is what we call a *supernova.* A burst of light, perhaps a billion times brighter than an ordinary star, appears in the sky where the star was located. It remains this bright for a matter of hours or days and then gradually dims over a period of a few months. This period is so short that supernovae are seen only rarely. When the explosion occurs in our region of the Milky Way galaxy, the supernova can be so bright that it is visible in the daytime. This has happened only a few times in recorded history, the best documented examples in 1054, 1572, and 1604. The supernova of 1054 was observed by Chinese astronomers and also recorded in cave paintings by American Indians. Supernovae are now frequently observed in other galaxies, where the event can be seen only with the aid of a telescope.

But remnants of the supernova of 1054 can be seen today. The Crab nebula is a glowing cloud of expanding gases (fig. 6.8). Scientists can measure today the rate at which these gases are expanding from a central point and can thereby calculate the time in the past when the gases began to expand. This time turns out to be about 900 years ago. Furthermore, the Crab is located in the same region of the sky as was the supernova, according to historical accounts.

Fig. 6.8. The Crab nebula, a remnant of a supernova explosion (Courtesy Palomar Observatory)

The story of the fate of burned-out stars is like a jigsaw puzzle. Diverse elements have to be fitted into place: the observations of white dwarfs, red giants, and pulsars; the early theory of the neutron star (see app. D); the theory and observation of supernovae; the Crab nebula. One more piece fits in to complete the puzzle. If we look in the direction of the Crab nebula, we find a pulsar, called the Crab pulsar. It is the core which remained after the supernova explosion of 1054. It pulses at a rate of $1/30$ of a second. Since pulsars slow down as time goes on, it is not surprising that this pulsar, having been formed only an instant ago in astronomical time, is one of the fastest.

Black Holes

There is one more card up the sleeve of the astrophysicist, something we have to take account of as we consider more and more massive stars. The white dwarf and the neutron star exist as stable objects because internal pressures (discussed in App. D) are able to counterbalance the inward attraction of gravity. But if a star is sufficiently massive, its gravitational force becomes irresistible. The star will collapse to a density greater than that of a neutron star and greater than that of the atomic nucleus; it becomes a *black hole.* As the star collapses, its gravitational attraction becomes greater and greater (since attraction increases when pieces of matter are closer together). It is an object, in other words, that is forever collapsing, and collapsing at a forever increasing rate.

The idea of the black hole is a consequence of Einstein's theory of general relativity—really a theory of gravity—and has been extensively developed by Subrahmanyan Chandrasekhar. But there is a way to think about black holes that goes back to the eighteenth-century French mathematician Laplace, based on the older Newtonian theory of gravity. A black hole is an object so dense that it creates around itself a region in which the gravitational force is so great that nothing can escape. Anything that enters that region is trapped and ultimately becomes part of the black hole itself.

The idea that something can be trapped by a gravitational force is not unfamiliar. We humans are trapped by the earth's gravitational field and cannot escape it without the aid of a rocket to accelerate us to high speeds. There is an "escape velocity" for the earth, a minimum speed that any object must have in order to escape the gravitational field and go off into space. (It is the velocity needed to travel as far from earth as we please, a velocity greater than that needed to reach earth orbit.) The escape velocity on earth is about 7 miles per second; on the moon it is only about 1.5 miles per second (it is smaller because the moon is less massive). The escape velocity depends on both the mass and size of an object, and we can imagine an object so dense that the escape velocity for it becomes very large and exceeds the speed of light. But nothing can go faster than the speed of light (see chap. 10), and so nothing can escape from such an object. This is the essence of a black hole. No light, no electromagnetic radiation of any kind, and no matter leave it. In principle, a black hole cannot be detected in any of the ordinary ways we detect things (as we see stars and planets and each other), by seeing the light emitted or reflected by them.

No conclusive evidence has been found that black holes exist in space. Yet there are indirect ways of detecting their presence, and many astronomers believe that one has been found. Stars are often found in the form of binaries. If one of two stars in a binary pair became a black hole, the two objects might continue in their orbits about each other. We would see a single star moving in an orbit, but there would be nothing visible about which it was orbiting. This in itself would not be unusual, because there are many examples of binaries in which one of the stars is too dim to be seen in the telescope. However, in the case of a black hole and a companion star, their close interaction might produce some characteristic signature of the black hole. Astrophysicists argue that filaments of matter might be pulled from the star by the black hole's intense gravitational force and in the process become heated to millions of degrees. This matter then would emit radiation in the x-ray region, some of which (outside the volume in which the black hole traps all radiation) proceeds out to space. Thus we look for a star and an x-ray source in mutual orbit. Further, by studying the orbit we can deduce the mass of the invisible companion. If this mass is significantly higher than 3 solar masses, it is a good bet that the unseen object is a black hole. A number of x-ray sources, notably one named Cygnus X-1 (because it is in the constellation Cygnus, the swan), are good candidates for black holes.

There are some loose ends in the theory of stars and their evolution. The most important is the lack of observed solar neutrinos. Detailed theories of supernovae and of the formation of pulsars have not been fully developed. The case of the black hole is still pending. It is nevertheless a profoundly impressive feat of modern astronomy to have accounted for all the major sources of light by constructing theories of the history of stars and by correlating these with the great variety of objects in the sky.

The Cosmos

To the scientist light is not only a messenger but also a tool, a tool of limitless versatility, which can be used to uncover the mysteries of the atom and of the stars. Nowhere is this more evident than in the fields of extragalactic astronomy and cosmology. As we attempt to understand the large-scale structure and behavior of the universe, we seek out fainter and fainter signals from deep space, we search the electromagnetic spectrum beyond the narrow confines of the visible, and we anlayze the subtle effects of matter and motion on light.

Beyond those stars in the neighborhood of the solar system we begin to find a hierarchical structure. Stars are collected into galaxies, galaxies are collected into clusters of galaxies, and the universe as a whole (as far we we understand it) partakes of a uniform expansion. Einstein wrote, "The most incomprehensible thing about the universe is that it is comprehensible." How is it that, solely by the gathering in of electromagnetic beams, we have come this far in comprehending the universe (admitting always that there are unknowns in front of us)?

For each advance into space one crucial step is prerequisite: the measurement of distance. For, as with the stars, if we do not know how far away an object is, we cannot determine its intrinsic brightness, and so we lose our chief probe into its nature. The method of parallax, as discussed in the previous chapter, can be used to measure distances to near stars. But parallactic motion is smaller the more distant the star is, and for objects at more than about 40 parsecs the parallax is too small to be seen. Astronomers have found that by analyzing what are called the *light curves* of certain stars—their variation in brightness over time—they are able to measure distances up to millions of parsecs and thus to establish, first, the structure of our own galaxy, and, later, the existence of other galaxies.

The Milky Way Galaxy

In the early twentieth century the entire universe was viewed as a disk—what is now called the galactic disk. It was a collection of stars, about 3,000 parsecs in diameter and 1,000 parsecs high,

Fig. 7.1. A portion of the Milky Way in the constellation Sagittarius (Courtesy Palomar Observatory)

Fig. 7.2. A globular star cluster in Tucana (Courtesy National Optical Astronomy Observatories)

with the sun in the center. It represented the extent of the known universe at that time (although we know now that this matter is all within our own galaxy). Farther than these distances there were no stars, although the power of existing telescopes would have been adequate to detect them. The asymmetry inherent in the disk is indeed visible to the naked eye. For in any direction in the plane of the disk (as opposed to along the axis of the disk) there is a greater concentration of stars, a cloudy band of brightness that we call the Milky Way (fig. 7.1). Even with a small telescope individual stars can be resolved in it, as was discovered by Galileo.

But outside the confines of the galactic disk lie about a hundred *globular clusters* (fig. 7.2), spherical associations containing tens or hundreds of thousands of stars, and these clusters provided the first evidence that the sun was not the center of the universe. Some clusters are as far away as 20,000 parsecs, and in 1917 the American astronomer Harlow Shapley (see fig. 6.3) found a way to determine these distances by observing a class of stars known as RR Lyrae stars.

These are variable stars; their brightness waxes and wanes regularly with a period of a few hours. Unlike main-sequence stars, which remain in equilibrium and constant in brightness for perhaps billions of years, variable stars are those caught in a brief period of instability. Observations of RR Lyrae stars relatively close to the sun had shown that they all had approximately the same intrinsic brightness. Shapley located such stars in globular clusters. Making the assumption that these stars in clusters had the same intrinsic brightness as the nearer ones, he was able to deduce the distance to the stars and hence to the globular clusters. In this way he determined that the clusters as a group form an approximately spherical system, across the center of which lies the galactic disk. Shapley correctly deduced that the system of clusters forms a rarefied outer haze around the disk; it is called the halo. His measurements give a diameter of about 30,000 parsecs to the disk of the galaxy, with the sun about 10,000 parsecs from the center (fig. 7.3). Thus was completed the work of Copernicus in removing humanity from the center of the universe.

At that time the word "galaxy" was used interchangeably with "universe," and what Shapley had done was to extend the size of the known universe by approximately a factor of ten. It was subsequently realized that the reason we cannot see farther along the galactic disk is that, distributed among the stars, are dust particles that absorb a considerable fraction of the passing light. This dust tends to absorb shorter wavelengths more effectively than long wavelengths so that there is a general "reddening" of starlight. Indeed we now make use of this observation to estimate how much interstellar dust there is.

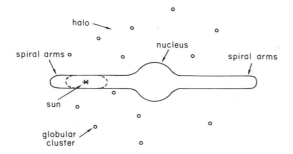

halo

nucleus

spiral arms

spiral arms

sun

globular
cluster

Fig. 7.3. A schematic view of
the Milky Way galaxy. The
dashed curve shows the early
twentieth-century view of the
galaxy, which was then considered to be the entire universe.
Astronomers have mapped the
nearby spiral arms of the galaxy by observing clouds of
neutral hydrogen gas, via their
21-centimeter radio emission.
The galactic center is observed
in infrared, since this radiation
is less strongly absorbed by intervening clouds of dust.

There is more to the galaxy than merely the stars and the
dust. Observations of some "stars," even under fairly low magnification, show that they are not points of light but clouds of luminosity called *nebulae*. Spectroscopic analysis of some nebulae
show characteristic lines of the emission spectrum of hydrogen.
These are the *emission nebulae* (fig. 7.4*a*), often quite beautiful,
dressed in the red of prominent hydrogen lines. The source of excitation of an emission nebula is a bright star embedded in the
cloud, one high along the Hertzsprung-Russell diagram, which
emits substantial ultraviolet. These energetic photons ionize the
free hydrogen atoms (separate the electrons from the nuclei),
creating a plasma, a cloud of protons and electrons swirling about
in space. Electrons are eventually captured into one of the highly
excited hydrogen states, and as they cascade down through various atomic levels the atom emits its characteristic lines in the
visible. In the constellation Orion, seen in the winter sky, there
are three closely spaced stars forming the giant's sword, and the
central one is an emission nebula. The four stars that provide its
excitation can be captured in a time-exposure photograph.

We use the term *HII region* to refer to interstellar hydrogen
in the state of a plasma. In contrast, an HI region consists of hydrogen gas in the form of neutral atoms. (Hydrogen on earth is
found in a third form, molecules made of two bound atoms.) Although HI regions do not emit visible light, they can be detected
in another way, which vividly illustrates modern astronomy's reliance on electromagnetic waves beyond the visible. The spectrum
of hydrogen is more complex than we have indicated so far. If we
look very closely at a single line in the line spectrum, we find it
has an internal structure; it is divided into a series of very closely
spaced narrow lines. This "splitting" of lines is a result of several
factors, but the one that concerns us here has to do with the *spin*
of electrons and protons (and is technically known as "hyperfine
structure"). Each of these elementary particles spins on an axis
much as the earth does. Spin is an inherent part of the particle's
existence; it cannot stop spinning. But it can change the direction

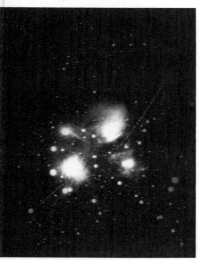

of its spin axis (as the earth does in precession). The total energy of the atom will be very slightly different depending on whether the electron and proton spins are antiparallel (the lower energy level) or parallel (the higher energy level). Therefore each orbital level (characterized by the electron's orbit) is split into two closely spaced spin levels. When an atom makes a transition from the upper spin level to the lower spin level, a very low energy quantum is emitted, corresponding to a frequency in the radio region; its wavelength is 21 centimeters.[1] This radiation has become one of astronomy's basic tools, as we shall see.

Sometimes, superimposed on the bright line spectrum of an HII region there is a continuous spectrum. And we sometimes find nebulae in which most of the luminosity lies in a continuous spectrum. These are called *reflection nebulae* (fig. 7.4*b*) and are understood to be the solid particles of interstellar dust, reflecting light of nearby stars. These nebulae tend to be blue, because it is largely the brightest (hottest) stars that illuminate neighboring dust clouds, and such stars have greater intensity toward the blue.

We also observe dust clouds by their absorption of light. If dust is located in front of an emission nebula it may absorb light from behind it, creating a dark pattern called an *absorption nebula* (fig. 7.4*c*).

The Galaxies

Not long after Shapley's delineation of the Milky Way galaxy (our galaxy) astronomers demonstrated the existence of other galaxies located at even greater distances. This advance was founded upon the light curves of another class of stars, known as *Cepheid variables*. These stars vary in brightness with periods ranging from a day up to a few weeks. In the course of studying nearby Cepheids

Fig. 7.4. Nebulae: (*top*) an emission nebula, one of the "stars" in Orion (Courtesy Mount Wilson and Las Campanas Observatories, Carnegie Institution of Washington); (*bottom*) reflection nebulae in the Pleiades (Courtesy Yerkes Observatory); (*right*) an absorption nebula, the Horsehead dark nebula in Orion (Courtesy Palomar Observatory)

(see fig. 7.8a), astronomer Henrietta Leavitt had established a one-to-one correspondence between the period and the intrinsic brightness (averaged over a period). The brighter stars, which are also the larger ones, oscillate more slowly. Thus it became possible to use Cepheid variables as a device to measure distance, determining first their intrinsic brightness and from that their distance. Cepheids are bright enough to be seen at distances of a few million parsecs.

The story of galaxies beyond our own goes back to the eighteenth century and the early telescopic observations of nebulae. In 1755 Immanuel Kant speculated that these clouds, if very distant, could in reality be collections of immense numbers of stars, "island universes" in his phrase, separated from ours by vast reaches of empty space. Thousands of nebulae were identified and catalogued (although their distances were not known) during the eighteenth and nineteenth centuries, notably by William Herschel and his son, John. In the early twentieth century a pattern began to be discerned. Nebulae in the plane of the galactic disk are irregularly shaped, resembling clouds of gas, while nebulae in a direction out of the plane are symmetrically shaped. Furthermore, whereas the former seemed to have bright line spectra, the nebulae out of the disk had continuous spectra with absorption lines, like stars. With the construction of new telescopes it became possible to resolve individual stars within the symmetrically shaped nebulae.

Definitive evidence for the existence of other galaxies was presented finally in 1924, when Edwin Hubble (fig. 7.5), at the Mount Wilson Observatory, studied Cepheid variables in several of the "nebulae" and thereby determined the distances to these systems. He found, for example, that the object we now know as the Andromeda galaxy (fig. 7.6) is close to one million parsecs (one *megaparsec*) from the solar system, a distance much greater than the dimensions of our galaxy.

To reach distances beyond the range of Cepheids, astronomers have used techniques similar to those using variable stars. Finding a class of brighter objects—certain blue supergiants, for example—believed to be all about the same intrinsic brightness, we can determine their intrinsic brightness by observing some of them that happen to be nearby. The brightness of one of these can be measured, for example, if it is near a Cepheid. Once its distance is thus known, its intrinsic brightness can then be calculated. Having made this calculation, we may now find blue supergiants at greater distances and, knowing their intrinsic brightness, determine how far away they are. We proceed in this way, leapfrog fashion, to measure distances to objects in galaxies farther and farther away—hence to measure distances to those galaxies: blue supergiants up to 25 megaparsecs, bright supernovae up to several thousand megaparsecs.

Making observations over these vast distances and over all directions in space, Hubble and others studied tens of thousands of galaxies, distributed more or less evenly over space. The galaxies fall into classes according to size and shape (fig. 7.7). Andromeda is an example of a large spiral galaxy, with a central bright nucleus and spiral arms twisting outward, like a spider caught in a whirlpool. The Milky Way galaxy is also a spiral, similar in size to Andromeda. Galaxies can also be elliptical or spherical. Some spiral galaxies have a bar of light across the center, with their spiral arms emerging from the ends of the bar. The spiral galaxies are generally the largest, containing some hundred billion stars. Elliptical galaxies are usually smaller, although there are some ranging to the very large and very small. By making radio observations of the 21-centimeter wavelength, we determine that hydrogen gas is located primarily in the arms of spiral galaxies, and not in their nuclei, and that hydrogen gas is not common in the ellipticals. Since free hydrogen gas is the source of new stars, we conclude that the spiral arms are the birthplace of stars.

Indeed, where we can determine the age of stars, we find the youngest stars in spiral arms.

Distribution of the galaxies in space is not entirely uniform. They tend to cluster, in groups of just a few galaxies or clusters of several hundred or several thousand galaxies. Our galaxy is a member of a small cluster, known prosaically as the Local Group. It contains approximately twenty galaxies (there may be others hidden by dust or too faint to be seen). Three are spirals, the Milky Way, Andromeda, and the small M33 (in the constellation of the Triangle); and the rest are small ellipticals or irregularly shaped galaxies. Noteworthy are the galaxies closest to us, the Clouds of Magellan (fig. 7.8), two irregulars that are visible only from the Southern Hemisphere (and were named by the crew of Magellan's ship, when the world was circumnavigated).

To finally comprehend the dynamics of the universe, scientists must know more than just how stars are collected into galaxies or how galaxies are distributed through space. They must measure motion, and from this motion unfold the forces that control matter at the grandest level. The measurement of motion in astronomy is based primarily upon one of the most important properties of light: the *Doppler effect,* first predicted by Johann Christian Doppler (fig. 7.9) in 1842. It applies not only to light but to any form of wave motion from a source that is moving toward or away from the observer. When the source is moving toward the observer, the successive wave fronts are crowded together by the approach of the source, so that the wavelength observed is decreased; it is smaller than it would be if the source

Fig. 7.6. The Andromeda galaxy. It is the nearest large spiral galaxy, a member of the Local Group. (Courtesy Palomar Observatory)

Fig. 7.7. Types of galaxies: (*left*) a spiral galaxy in Virgo; (*middle*) a barred spiral galaxy in Eridanus; (*right*) an elliptical galaxy in Andromeda. (Courtesy Palomar Observatory)

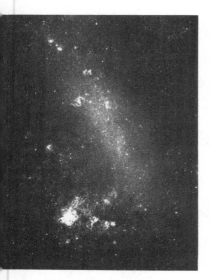

Fig. 7.8. (*top*) The Large Magellanic Cloud; (*bottom*) the Small Magellanic Cloud. Henrietta Leavitt established the relation between period and brightness of Cepheid variables by observing these stars in the Magellanic Clouds. (Courtesy National Optical Astronomy Observatories)

were stationary. This is called a "blueshift." Conversely, if the source moves away from the observer, the wavelength becomes longer: a "redshift." Moreover, the amount of this shift is proportional to the speed of the moving source. So if we know what the wavelength would have been for a stationary source, we can use the observed Doppler shift (red or blue) as a measure of the speed of the source.

In this way astronomers have used the Doppler effect as a tool to approach a wide variety of important problems: (1) to determine the rate of rotation of the planet Mercury by reflecting a radio beam from the surface and noting the Doppler shift of the beams reflected from the two sides, one moving toward us, one moving away; (2) to measure the period with which binary stars orbit one another, by observing the shift in the positions of spectral lines in emitted starlight; (3) to measure the motions of the stars near to the sun; (4) to detect the motion of nearby spiral arms of our own galaxy, by observing the wavelength shift in the 21-centimeter radio line; and (5) to determine the overall rotation of the Milky Way by measuring the rotation rates of stars in their orbits around the galactic center.

The Doppler effect also gives us a view of the universe at the largest scale, beyond our own galaxy and beyond our own cluster. The Local Group spans a distance of about 1 megaparsec, and at distances of 10 to 100 megaparsecs, we begin to see the clustering of other galaxies (fig. 7.10). Some clusters, like ours, contain a small number of galaxies; others may contain thousands. Within clusters the component galaxies move in complex orbits under the influence of their mutual gravitational attraction, much as the stars do within a galaxy. The clusters are so distant that we cannot detect angular motion of galaxies across the sky; but the component of motion toward or away from us is given directly by the Doppler shift in the wavelengths of bright hydrogen lines (emitted by HII regions in these galaxies). If we assume that the law of gravitation acting at this cosmic scale is the same as that deduced from observations on earth and in the solar system, a strange inconsistency emerges. Given the orbits of galaxies in a cluster, we can deduce an approximate value for the total mass in the cluster. And given this, we calculate the ratio of mass to total luminosity (rate of light emission) for the cluster. This mass-luminosity ratio turns out to be considerably larger than the corresponding ratio for a typical nearby single galaxy or for a typical star. In other words, there appears to be *invisible mass,* a substantial quantity of matter (perhaps 90 percent or more of the total) that is in some form other than the usual kinds of stars and that does not produce light. More recently evidence for invisible mass has also been uncovered in the outer regions of single galaxies.

What form this dark matter takes remains a mystery. We are

Fig. 7.9. Johann Christian Doppler (Courtesy AIP, Niels Bohr Library)

familiar with various astronomical objects that produce no light (planets and burned-out stars, for example), but none is known to make up a significant part of the mass of the Milky Way or nearby galaxies. There may be mass in the form of black holes; or in the form of neutrinos, if neutrinos have mass; or in other still more exotic forms. Whatever the eventual solution to the puzzle of invisible mass, the centuries-old habit of regarding light as a measure of matter, and the stars as all, no longer survives.

The Expansion of the Universe

We move, finally, beyond the scale of clusters of galaxies to the universe as a whole, still using the Doppler effect as a guide. And we reach the cosmic redshift, discovered by Hubble in 1929. Galaxies in clusters, including the galaxies in the Local Group, move about one another in a local dance. But if we take a larger view and determine the motions of galaxies in many directions at distances up to 50 or 100 megaparsecs, we find consistently a shift of spectral lines toward the red. More important, there is a direct proportionality, known as *Hubble's law,* between the distance to a galaxy and the degree of the redshift. Barring local fluctuations, the galaxies all seem to be moving away from our galaxy, and those that are farther away are moving away at greater velocities (see app. E). For a distance of 100 megaparsecs, the velocity of

Fig. 7.10. A cluster of galaxies in Hercules (Courtesy National Optical Astronomy Observatories)

recession is about 5,000 kilometers per second. Hubble's law is established for those galaxies near enough that their distance can be measured. It can be verified up to distances greater than 1,000 megaparsecs, although distance measurement becomes less and less reliable as we look farther away. Beyond a certain point we have no independent means of measuring distance and so cannot verify the proportionality between redshift and distance. Instead scientists have assumed the validity of Hubble's law for extremely large redshifts and on this basis have determined distances in the range of several billions of parsecs. In this realm we find the most mysterious and controversial objects in the heavens, the *quasars*.

The discovery of quasars (quasi-stellar radio sources) resulted from a combination of research in the visible part of the spectrum and in the radio region. Radio astronomy has been a particularly important and active field in the last three decades. Radio studies have shown, among other things, the presence of matter other than hydrogen atoms in the Milky Way (molecules of H_2O, for example), the existence of weak radio sources at the center of our galaxy as well as of many of the other "normal" galaxies, the existence of "radio galaxies," which emit much more of their energy in radio than in the visible and which seem, in some way, to be undergoing violent or explosive processes.

What has made radio observation an effective tool and led to the discovery of quasars is the construction of radio telescopes with high resolving power. Remember that what determines the

Fig. 7.11. The 300-foot movable radio telescope at Green Bank, West Virginia. The large metal dish can be pointed at different directions in the sky and focuses radio waves onto a receiver by reflection. The dish is shaped like a concave mirror, the kind used in a reflecting optical telescope. The radio reflector does not have a highly polished, extremely smooth surface, as does a mirror, since specular reflection requires that the surface be smooth only over a distance approximately equal to the wavelength. (Courtesy National Radio Astronomy Observatory, Associated Universities, Inc.)

Fig. 7.12. The Very Large Array (VLA) radio telescope near Socorro, New Mexico. To obtain better resolution of radio signals, a sophisticated method has been devised whereby two or more radio telescopes, separated by a large distance, can be focused upon the same object; the signals are combined in such a way that the resolution is equivalent to that of a single telescope whose diameter is the distance between the two. This method, known as *interferometry,* involves precise correlation of the two signals by means of an atomic clock and their synthesis in a computer. The VLA consists of twenty-seven radio telescopes, each 25 meters in diameter, deployed in the shape of a Y. The telescopes along each arm can be moved on railroad tracks to separations up to 21 kilometers. The array can achieve resolutions similar to those of the best optical telescopes. Astronomers have also combined signals from two very distant points on the earth (California and Australia in one experiment) to obtain a resolution corresponding to a telescope diameter of thousands of miles. Plans are being made for a new radio instrument, the Very Long Baseline Array, which will consist of ten receivers in Hawaii, Puerto Rico, and the continental United States. Scheduled for completion in 1990, the VLBA will have a resolution about 1,000 times better than the VLA. It will allow detailed studies of the structure of quasars and other compact sources of radio energy. (Courtesy Radio Astronomy Observatory, Associated Universities, Inc.)

ability of a telescope (optical or radio) to resolve two objects that have a small angular separation or to map out the structure of a small object is the ratio of wavelength to the diameter of the objective lens or mirror. In the case of light, with wavelengths of some hundred thousandths of a centimeter, we attain excellent resolution with telescope mirror diameters of a meter or less. For radio signals, with wavelengths of a few centimeters, we need a radio telescope a hundred thousand meters (a hundred kilometers) in diameter to attain the same resolution. Thus the major drive in radio astronomy has been to search for ways to produce large-diameter receivers (figs. 7.11 and 7.12).

Using these devices, radio astronomers have been able to map the radio structure of many areas of the sky and plot the shape of various intense radio sources. By 1960 they began to correlate the positions of radio sources with known optical sources. In many cases the optical counterpart of a radio source appears as a normal galaxy or as a galaxy exhibiting peculiarities in shape, indicating some violent activity. But some of these optical structures turned out to be very small and to look, on telescope photographs, simply like ordinary stars. Pursuing the subject further, in 1962 Maarten Schmidt studied the spectrum of one of these objects and found it to have a series of bright lines which at first seemed unlike those of any known element. Within a short time, however, he realized that the line spectrum was nothing other than the spectrum of the hydrogen atom with the lines shifted to the red, but shifted to a much greater extent than is the case for the galaxies studied by Hubble and his followers. This was the discovery of the first quasar (many hundreds of others have since been found, with similar properties) and the origin of their mystery. For according to Hubble's law, which established a correlation between redshift and distance, quasars are the most distant objects yet seen. They are receding from us at speeds that are a considerable fraction (80 or 90 percent in some cases) of the speed of light. Although some have been found with smaller redshifts, statistical analysis shows that the great majority have large redshifts and are located in the farthest reaches of the universe, at distances of billions of parsecs. Thus, although its optical appearance may be that of a dim star, the great distance of the quasar implies that it produces a tremendous quantity of energy. In the visible it may produce as much light as a hundred galaxies like ours; in the radio it is the brightest object in the sky (fig. 7.13).

Moreover, this energy flows from a source that measures something like one-tenth of a parsec across, more than 100,000 times smaller than an ordinary galaxy.[2] Numerous quasar models have been put forth to explain the production of such prodigious quantities of energy in such a small volume: a dense collection of large stars, many exploding as supernovae; a giant black hole in the center of a galaxy, swallowing and disrupting surrounding matter; a single dense rotating object, something like a pulsar on a gigantic scale; matter and antimatter in collision. The solution to the mystery of the quasar remains one of the leading challenges of modern science.

It is important not only because quasars are peculiar objects in the sky but because they are located at the greatest distances from us. Since light travels at a finite velocity, the radiation received from quasars was produced in the distant past, many bil-

Fig. 7.13. The quasar 3C 273 (Courtesy High Energy Astrophysics Division, Harvard/ Smithsonian Center for Astrophysics)

3C 273
HRI 20 ARC SEC ⊢—⊣
1-4 COUNT CELLS ZEROED
EINSTEIN OBSERVATORY

lions of years ago. We begin to probe the structure of the universe as it was in the early stages of its existence. We question whether quasars, seen more frequently at great distances than nearby (as are radio galaxies also), represent some early stage in the history of ordinary galaxies, although we do not know how these turbulent objects might have evolved into galaxies. To pursue finally the history and structure of the universe on the largest scale, we return to the subject of Hubble's law, and the expansion of the universe.

The Big Bang

The redshift tells us that the galaxies are moving away from us, although we should not view it as if our galaxy is at the center of the universe. Rather, the redshift measures relative velocities, and we conclude only that every galaxy is separating from every other galaxy. What do we make of the observation that those galaxies that are farther away are the ones that are moving faster?[3] The answer is perhaps more apparent if we ask, why are the galaxies that are moving faster farther away? If at some time in the past the galaxies (or some more primitive structures which became the galaxies) were close together and were separating with various velocities, then, by the time of the present, the ones that originally had higher velocities would have reached greater distances. This is a reasonable inference from Hubble's law. But of course we must go beyond this and ask, where did the original velocities come from, particularly since, on the cosmological scale, the only force that we know of is the gravitational force, which is attractive. The attractive force, acting alone on the densely packed matter of the early universe, would have caused it simply to collapse, to collapse into a single cosmological black hole. That this did not happen must be due to the occurrence of a cataclysmic, explosive event, its ultimate cause unknown, which we call the *big bang*. Extrapolating backward in time from the measured expansion of the universe, we picture times when the universe was smaller, with matter at unimaginably high densities and temperatures, reaching finally the big bang some 15–20 billion years ago. In this way, despite some uncertainty in cosmological theory (which we use to extrapolate in time), we determine the "age of the universe." Furthermore we can independently calculate the ages of stars, using the theory of stellar evolution; and we find that the birth of the oldest stars goes back as far as, but not prior to, the big bang, thus corroborating the entire picture.

The expanding universe is not to be understood as a cloud of matter expanding into empty space—like the ink of an octopus expanding into the sea. Rather, the astronomer views matter as filling all space; however compressed matter was in the early times after the big bang, there still was no empty space surrounding it. The expansion of the universe means that the distances between the galaxies increase as time advances, no more, no less; or, put another way, the *density* of matter was large in the past and is decreasing. Space itself may be finite or infinite; that is a different, and unanswered, question, independent of Hubble's law.

The big bang cosmology also provides a solution to Olbers' paradox. Olbers argued that if the stars were uniformly distributed throughout infinite space, then the night sky would be as bright as the surface of a star or the sun. This conclusion requires that out in the night, no matter how far away we look, there will be stars brightly shining. But at great distances we are looking into the distant past. And there is a time in the past, before the big bang, to which we cannot look; or at least we cannot assume the existence of brightly shining stars and a universe like the one in which we live. Olbers' assumption of brightly shining stars at *all* distances simply cannot be made. Notice, further, that this resolution applies equally well to any universe of finite *age,* given the finite velocity of light. For the amount of light reaching us on earth does not depend on the total number of stars but on the number of stars whose light has had time to reach us. It is then a straightforward calculation to determine whether that (finite) number of stars is or is not enough to fill the sky. Whether one uses an age of 5,000 years (as might have been done by Olbers) or 15 billion years, one finds that the stars do not fill the sky; the night sky indeed should be dark.

In recent years theoretical physicists, armed with data from the new generation of particle accelerators and with modern theories of elementary particles, have had the temerity to construct theories of the early stages of the big bang, back to the epochal minutes and seconds after time began. We cannot linger over these speculations. But we must not omit discussion of one residue of the big bang which we believe has been detected in modern radio experiments and which has lent substantial support to the theory.

The early universe was a seething cloud of matter and electromagnetic radiation, so densely filled with charged nuclei and electrons that radiation could not travel far without interacting with matter; the universe was opaque. As it expanded and cooled, a time came when the temperature was low enough (a few thousand degrees) to permit the combination of nuclei and electrons—the first atoms. At this point, about a million years after the beginning, the interaction of radiation with matter became far weaker; the universe became largely transparent. Radiation became free to fly chaotically across the empty space of the expanding universe. We may picture the entire cosmos as a single blackbody, glowing uniformly with visible radiation, peaked near the red. What if we wish to see this radiation?

Fig. 7.14.
Robert Wilson
(*left*) and Arno
Penzias (Bell
Telephone
Laboratories;
courtesy AIP,
Niels Bohr
Library)

It is a "cosmic background radiation" and so comes from all directions in space. We look out toward objects at the greatest distances from us. Thus we receive light that has traveled the greatest time, photons created in that ancient glowing universe. At these great distances, as Hubble's law implies, we see light with its wavelengths increased by extreme amounts—the redshift. The original photons (with wavelengths around 10^{-6} meters) are shifted to wavelengths around 1 millimeter, radiation in the radio. This radiation was discovered in 1965 by Arno Penzias and Robert Wilson (fig. 7.14). It follows a blackbody curve for radiation corresponding to a temperature of 3 degrees above absolute zero,[4] in good agreement with theory. Most scientists consider it a direct confirmation of the theory of the big bang.

And what of the future? Will the universe continue to expand indefinitely? Or will the still powerful force of gravity eventually slow down the expansion of the galaxies, to the point where matter will reach a state of maximum extension and, like a yo-yo at the end of its string, return to the hand that threw it forth? Can it be that we find ourselves in the expanding stage of a universe that will later begin to contract? And will it then collapse in what some call the "big crunch," perhaps later to expand again after a new big bang? We do not know the answers to these questions, although it is perhaps easier for us to accept the oscillating universe, extending indefinitely backward and forward in time, rather than a single big bang, before which there was no existence and after which there is infinite dissolution.

Yet a scientific answer may be found. Since we know how rapidly the universe is expanding, theory allows us to calculate just how much matter there must be to arrest the expansion. Specifically, there is a "critical density" (corresponding to a few atoms per cubic meter, on the average), such that if the actual density of the universe is greater than the critical density, then expansion will stop and reverse; while if the actual density is less than the critical density, then expansion will continue forever. What is the actual density of the universe? If we count galaxies and average their mass over the great regions of empty space which they occupy, we find the density of matter to be considerably smaller than (about one-hundredth of) the critical density; there is not enough matter to stop the expansion. But counting galaxies is not enough, for we know now that there is invisible matter, making its presence felt, not in light from galaxies, but only in gravitational effects. Our best estimates of the quantity of dark matter, based on the motions in clusters of galaxies, imply a total density of the universe which is still only about one-tenth of the critical density. Both these numbers are, of course, known only with some uncertainty, and it does seem strange, surprisingly coincidental, that the universe should be so close to the borderline between eventual collapse and perpetual expansion. Indeed, recent speculations about the earliest moments of expansion suggest that the actual density should be precisely equal to the critical density (implying a universe that expands indefinitely, with expansion velocities that approach zero as time goes on). If this were the case, there would have to be some new form of invisible mass, beyond that already hypothesized for galactic clusters. Correspondingly, recent theories of elementary particles propose the existence of strange new particles that might make up this invisible mass. It is too soon to speculate on where these hypotheses may lead.

For another approach to the problem of the expansion of the universe, consider the galaxies in the early moments of their expansion, when matter was closer together and the effects of gravity were immensely greater. The proportionality between distance and velocity embodied in Hubble's law is understood if we make the assumption that galaxies move at constant speed. But in the early stages the assumption would be far from valid; galaxies must surely have been decelerated by their mutual gravitational attraction. Thus Hubble's proportionality would fail to hold. The degree of failure of Hubble's law at early times would tell us how strong gravitation really was, and thus whether the universe will expand indefinitely or not.

We look back in time by looking at the most distant galaxies. We measure their redshifts, thus their velocities, and try to determine independently (that is, without using Hubble's law) how far they are from us. Thus we attempt to see to what extent and in what way Hubble's proportionality fails. Here finally the astronomer is overwhelmed by uncertainties—uncertainties in the measure of distance and uncertainties about how the brightness of galaxies may change over time. Again the question of the future remains open, but it is not beyond the realm of possibility that light from the greatest distances will, before too long, provide an answer.

Light and Life

8

"The night was pitch dark, with a fresh breeze. The sea from its extreme luminousness presented a wonderful and most beautiful appearance; every part of the water which by day is seen as foam, glowed with a pale light. The vessel drove before her bows two billows of liquid phosphorus, and in her wake was a milky train. As far as the eye reached the crest of every wave was bright; and from the reflected light, the sky just above the horizon was not so utterly dark as the rest of the Heavens. It was impossible to behold this plain of matter, as it were melted and consumed by heat, without being reminded of Milton's description of the regions of Chaos and Anarchy."[1] So wrote the youthful Charles Darwin of his voyage in the South Atlantic on the HMS *Beagle* in 1832. Had Darwin taken a flask of seawater he might have observed floating in it the one-celled bodies of *Noctiluca miliaris,* each visible to the naked eye, one of nature's most ubiquitous bioluminescent forms. The sun rules the day, the moon and stars rule the night, but in the depths of the sea, across the surfaces of fields and waves, with gentle rearrangement of molecules living things create their own light.

The word *luminescence* is used in contrast to *incandescence,* which refers to light emitted as a consequence of high temperature. The sun, the stars, and the electric light bulb are examples of incandescence. The gas discharge tube, the fluorescent bulb, and the aurora are examples of luminescence. Another instance is *chemiluminescence,* in which light is emitted as a consequence of a chemical reaction in which one of the reaction products has been left in an excited state. (See, for example, the discussion of the airglow in chap. 5.) The element phosphorus gives off a faint glow as a result of its spontaneous chemical reactions with oxygen in the air.[2] Other examples of chemiluminescence are the oxidation of sodium and potassium, the reaction of chlorine with vapors of these metals, and the luminescence of certain living organisms. The "burning of the sea" (also called phosphorescence in the older literature) has been observed by travelers and philosophers as far back as 500 B.C. It is seen at night in many places, particularly in tropical waters, when the sea is disturbed by the prow of a ship or the passing of a fish. In the eighteenth century it was recognized to be due to the flashing of protozoa.

Bioluminescence

Study of the luminescence of living things takes us far and wide across the plant and animal kingdoms. Although luminous species (figs. 8.1 and 8.2) are broadly distributed among the phyla and classes—about half of all phyla contain examples—the number of luminous species is an extremely small fraction of the total number of known species. Luminosity often has adaptive value, but for different types of creatures it seems to have quite different functions, and for some it has no function at all. The distribution over the plant and animal kingdoms seems to be random, so that, with certain exceptions, we do not see clear evolutionary lines as we move up the phylogenetic scale.

Numerous kinds of bacteria and protozoa are luminous, as well as certain species of fungi; no examples are found among the higher plants. Among the higher animals, the greatest number of luminescent species is found in the phylum Arthropoda, particularly among the insects (the firefly is the best known example) but also among various crustaceans. In the sea many jellyfish and comb jellies luminesce, and many squid and deep-sea fishes have highly developed luminous organs. Among the higher vertebrates—amphibians, reptiles, birds, and mammals—bioluminescence is absent.

Among marine animals luminescence can take the form of a secretion into the sea from luminescent cells. These cells may be

Fig. 8.1. *Mnemiopsis* is a small invertebrate found in waters off the East Coast of the United States. (About twice natural size.) When disturbed at night, the animals luminesce along the ciliated plates shown in the photograph. (Photograph by Ralph Buchsbaum, *Animals Without Backbones,* 2d. ed. [Chicago: University of Chicago Press, 1948], 108-2.)

Fig. 8.2. Sea pens are luminous animals that grow in colonies under the sea. The branching parts (polyps) are purple and glow a bluish-green when stimulated. (Courtesy Frank H. Johnson)

scattered over the surface of the animal's body, concentrated in certain areas, or collected into a structure that we would call a luminous gland. Such variations parallel what we find for *photosensitive* cells, the presumed precursors of eyes. Moreover, higher animals such as squid and fish may have intricate light-producing organs, which parallel the highly developed light-receiving organs evolved in these creatures. Light production by these animals is intracellular (that is, luminous material is not secreted but remains in the producing cell) and associated with complex structures for light control (lenses, reflectors, and opaque and colored screens).

Some forms—bacteria, fungi, and certain fish—luminesce continuously, but most do so only when stimulated. Stimulation can be internal, controlled by the animal's nervous system or hormonal system, or external, influenced by the physical environment (the flashing of dinoflagellates when disturbed by the passing of a fish). For some species luminescence is periodic, in the sense that they flash only at night and cannot be stimulated to do so during the day. Some are seasonally periodic and luminesce only during spring or summer at the time of breeding.

Since ancient times it has been known that decaying meat and fish can emit light, but it was not recognized until about 1875 that this luminescence was not due to the dead animal itself but rather to luminescent bacteria (fig. 8.3). (This was about the same time that the bacterial origin of disease was demonstrated.) Luminescent bacteria generally do not thrive at the elevated body temperatures of warm-blooded creatures, and so they are not pathogenic to humans. In 1887 P. Tollhausen sprinkled his dinner with a luminescent culture of *Bacterium phosphorescens* and dined thus for three nights, verifying that they were not harmful. It should be

Fig. 8.3. Flasks filled with
dense cultures of luminous
bacteria. The photograph
is exposed by the light of
the bacteria. (Courtesy
Frank H. Johnson)

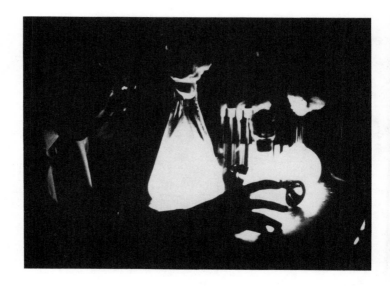

Fig. 8.3. Flasks filled with dense cultures of luminous bacteria. The photograph is exposed by the light of the bacteria. (Courtesy Frank H. Johnson)

noted that this courageous man had first injected the culture into his guinea pigs and also fed it to his cat. While some luminous bacteria live on dead organic matter, other strains are parasitic, causing infection or death in various cold-blooded animals. Other strains live successfully in or on healthy animals, providing at the same time a service to the animal. Indeed, sometimes that service is the light itself. For certain fish and squid possessing a light organ, the light source is symbiotic bacteria, and the animals have evolved their own lenses and reflectors to focus that light and utilize it. Other species have similar organs with their own light-producing cells. On a number of occasions scientists have mistakenly assumed an organism to be self-luminous. E. N. Harvey once observed a frog with a luminous belly, only to discover that the frog had just made its dinner of fireflies.

Despite the variety of their habitats, luminous bacteria do not seem to differ in any fundamental way from each other or from nonluminous strains. In fact their luminosity is not crucial for their survival; nonluminous mutant strains are easily produced and live quite successfully.

Similar luminescence can be observed on rotting wood, on the freshly exposed surfaces of logs turned in a forest at night. This is due to certain fungi that are luminous in the presence of air. Both bacteria and fungi will fade if deprived of oxygen and then will glow more brightly than usual when supplied again with oxygen. Thus it is suggested that bioluminescence is a chemical process involving oxidation, similar to ordinary chemical burning. When deprived of oxygen the organism continues to produce fuel, which accumulates, so that it burns more brightly when oxygen again becomes available. The chemical nature of the process is also indicated by the fact that luminosity varies with temperature. At about $-10°C$ the light of bacteria disappears (although the organism can survive down to much lower temperatures), and, within limits, luminosity increases as temperature increases.

In the sea luminescent forms of microscopic life—radiolaria, and particularly dinoflagellates—are abundant from the Pacific coast of the northern United States to the Mediterranean Sea, to the Indian Ocean and to northern Asia. The red tides of the sea are due to dinoflagellates, as can be patches of other colors, brown and yellow, seen in the day. These colors are not luminous, but at night the organisms glow with pale blue light. Bright points of light in the water are likely to be due to jellyfish, or similar creatures, but a homogeneous area of luminescence indicates the presence of dinoflagellates. *Noctiluca miliaris* is a famous example of a luminescent dinoflagellate, easily visible to the naked eye, with granules of photogenic material in its protoplasm. *Noctiluca* responds during the night even if kept in a bottle in artificial light and does not respond during the day even if kept in the dark. This day-night periodicity makes it seem likely that luminescence has some survival value, although it is difficult to say precisely what it is. In contrast, the constant luminescence of some bacteria seems to be quite irrelevant to their survival and is believed to be merely a vestige of some earlier function.

The variety of luminescent marine forms is endless (although freshwater species are rare). The boring clam, *Pholas dactylus,* provided science with important insight into the nature of bioluminescence. Raphael Dubois, in 1887, obtained a luminous extract of *Pholas* and isolated from it two substances, an enzyme, which he called *luciferase,* and a fuel, which he called *luciferin.* The enzyme catalyzes a chemical reaction in which luciferin is oxidized in air. The chemical reaction leaves the product molecules in an excited state, which subsequently decays with the emission of light.

The crustacean *Cypridina hilgendorfi,* from waters near Japan, was studied extensively by Harvey. *Cypridina* secretes into the water two kinds of granules, one of luciferin and one of luciferase,[3] and these dissolve, emitting a blue luminescence. During World War II, Japanese soldiers used *Cypridina* as a source of light. A dried extract on the hand, moistened, emits light sufficient for reading. Attempts have been made at various times to use luminescent creatures as a source of permanent artificial light. The Banda islanders have used the luminous organ of a fish as bait in fishing. A jar of water containing the crushed bodies of several large jellyfish serves well as a lantern. Tropical fireflies in a glass have been used for ornamentation in South America and the Caribbean.

Uses of Luminescence

Of what use is luminescence to the animal itself? The firefly (fig. 8.4) and various other beetles use light as a mating signal. In a typical behavior pattern, the male flies above the grass at dusk and emits a characteristic signal, flashing his light on and off in a pattern recognized by the female, who lies in wait in the grass (females of some species cannot fly). She emits a flash in response, and after a number of further signals, the male flies down and they mate. It has recently been observed that some fireflies learn to imitate the signals of another species, so that, for example, a female may lure a male of another species to an expected sexual encounter, only to use him for her dinner. The larvae of many species are also luminescent, and these are called glowworms.

The light organs of the beetles are among the simplest in structure, consisting of a simple mass of photogenic cells. These are richly supplied with oxygen by a complex system of tracheal tubes, and this level of oxygen supply is the means by which the animal controls its light production. The firefly's flash is indeed "cold light," having a range of wavelengths centered on 560 nm (yellow-green) and entirely within the visible region. The peak of the spectrum varies from species to species by a rather small amount, up to 10 nm. Nearly all the colors emitted in bioluminescence are in the yellow-green to green to blue. (An exception is the beetle *Phrixothrix* of Central and South America. Its larva has

Fig. 8.4. Anatomical drawing of a firefly, from a plate published in the *Philosophical Transactions* in 1685 (Courtesy The Royal Society of London)

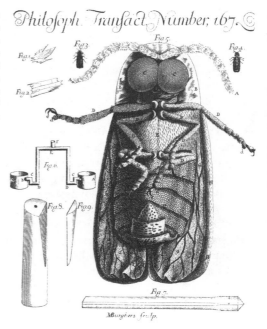

two luminescent red spots on the head and a row of eleven green lights on each side of the body. Called the "railroad worm," it keeps its red headlights on at night and flashes its sidelights when disturbed.)

W. D. McElroy and H. H. Seliger have studied the biochemistry of the firefly in detail, have obtained its luciferase and luciferin in pure crystalline form, and have determined the chemical structure of the latter. They have also shown that in the firefly (although this is not true for *Cypridina* or necessarily for other genera) a necessary ingredient for luminescence is adenosine triphosphate (ATP), a chemical known to be essential for many energy-related biological functions. This compound, which we will mention again, consists of adenosine (an element of genetic material) to which three phosphate groups have been added. (The phosphate group is a combination of a phosphorus atom with three oxygen atoms.) The addition of each phosphate group—particularly the last two—dramatically increases the energy of the molecule. Consequently, ATP serves as the energy-carrying system for many physiological and biochemical activities. To release energy it gives up one or two of its phosphate groups, to become adenosine diphosphate (ADP) or adenosine monophosphate (AMP), respectively. In the oxidation of luciferin to produce firefly light, ATP is necessary. It loses its energy and becomes AMP, to drive the production of light. (Indeed, firefly luciferin and luciferase can be applied to cells of other organisms, and the quantity of light produced used as a sensitive measure of ATP concentration. This is one of a number of ways in which luminescence has become an important tool in the biological laboratory.)

Marine worms observed in Bermuda and other areas of the Atlantic also use luminescence in mating. In the words of a traveler of 1911: "Quite suddenly [the female] becomes acutely phosphorescent, particularly in the posterior three-fourths of the body, although all the segments seem to be luminous to some degree. At this phase she swims rapidly through the water in small luminous circles two or more inches in diameter. Around this small vivid circle is a halo of phosphorescence, growing dimmer peripherally. . . . [Then the males appear], first as a delicate glint of light, possibly as much as ten to fifteen feet from the luminous female. . . . They rotate together in somewhat wider circles, scattering eggs and sperm in the water."[4] Thus the next generation. These creatures, also called fireworms, may have been the candle-like glints of light reported by Christopher Columbus on his first trip to the Bahamas.

Far from the coasts and deeper into the oceanic waters we find fish and squid with highly developed light-producing organs. A concentrated mass of photogenic cells is wrapped on sides and back by reflectors to turn all the light outward. Behind this structure there may be a layer of pigment cells, and in front a lens, similar to that of the eye, to focus a light beam out into the water.

Fish and squid display their light only in response to stimulation. Some fish produce light continuously and possess a screen that can be moved across the front of the light organ to darken it. One species rotates the entire organ downward or upward, to turn it off or on. Entering the organ from behind is an extensive system of blood vessels and nerves. In at least one case scientists have established (by injection of adrenaline) that the fish's light production is controlled by hormones.

The location of the light organs is different for each species. They may be on the sides, bottom, or the head of a fish or on the head, mantle, or tentacles of a squid. Location near the eyes is rather common, indicating a function as a lantern. In some cases there are many light organs, sometimes hundreds, so arranged that it appears that lights in different parts of the body have separate controls and separate functions. Little research has been done on the utility of lights in fish and squid behavior. Besides use for sexual signaling and for lighting the deep waters for predation, lights may be used for recognition among members of a school, perhaps even to frighten predators. One anglerfish bears a luminous organ on the tip of an appendage and uses it to lure prey into its mouth.

Yet despite the obvious value of luminescence to some species, it seems very possible that the functions to which light sources have been put are in some way secondary adaptations. Luminescence appears only randomly among the species, its existence is not crucial for many of the forms that possess it, and in some cases it appears to have no survival value. It may be, therefore, that luminescence played a more basic and widespread part in the evolution of life at a more primitive stage and later became vestigial in many living things, some of which then adapted it for other purposes.

The Absorption of Light and Ultraviolet

If the emission of light by living organisms has become secondary, or peripheral, in modern times, the *absorption* of light is central to life and has always been so. Ultraviolet light is very possibly the energy source that acted in the pre-biotic earth to form complex organic compounds from simple ones. We have seen that in the laboratory ultraviolet can spontaneously generate chemical building blocks of life when it is passed through gases modeling the primitive atmosphere. At the time of the transition from nonlife to life, an extensive variety of chemical "experiments" would have been carried out randomly in response to the sun's high-energy photons. Thereupon a form of "natural selection" may have acted—as some modern experiments suggest—to enhance the growth of molecules that were most effective in copying themselves, just as Darwinian natural selection encourages species that reproduce efficiently.

But as ultraviolet creates, so it can kill. Proteins absorb strongly at about 280 nm; nuclei acids at about 260 nm. Radiation in the latter region is the most lethal for bacteria, as well as the most effective in producing mutations and in retarding cell division. Ultraviolet modifies the bases in DNA, disrupting an organism's ability to grow. Primitive self-replicating systems would have had to collect some distance below the surface of water—or possibly in other protected environments—in order to survive (or else develop a mechanism to repair ultraviolet damage). Eventually organisms capable of photosynthesis evolved, producing oxygen and finally the ozone layer in the upper atmosphere. Then the earth's surface was protected from most radiation at wavelengths below 320 nm, and the way was cleared for the evolution of life on land.

Radiation reaching the earth today is confined to the region between 320 and 1,100 nm, the longer infrared being absorbed by water vapor and other gases in the air. It is a broad distribution with a maximum flux of photons at around 680 nm. (Above the atmosphere the maximum is at 636 nm.) It is important to note that the figure quoted here is not the wavelength of maximum intensity, which means maximum radiant energy, but the wavelength of maximum number of photons. (Maximum intensity above the atmosphere is below 500 nm.) Since each of the longer-wavelength photons carries less energy, the peak number of photons will be found at a longer wavelength than the peak for intensity.

We stress the photon count because in biology, as in physics, it is the individual photon that matters. Each photobiological event involves the absorption of a single photon by a single molecule, at least as the primary step. A single ultraviolet photon will kill a bacterium, if it hits the right spot. No quantity of red photons will. In the case of vision, while the human eye cannot detect a single photon, the primary process is the isomerization of retinal by one photon, and experiments have shown that as few as five or six photons can be processed by the retina and the brain into a visible flash of light. A third illustration, which we will come to shortly, is the absorption spectrum of chlorophyll, which has a maximum at 680 nm, just matching the peak photon count in the incoming radiation. The laws of quantum mechanics are often said to apply only to the world of the atom; but they cannot be ignored in the making of a tree.

The absorption of light, whether by chlorophyll in a leaf or by vapor in the air, depends on the energy level structure of the absorbing molecule. We have seen (chap. 5) that there will be a series of electron states, or levels (just as in the case of a single atom), and that built upon each electron level will be a series of closely spaced vibrational and rotational levels. These latter correspond to degrees of vibration and rotation among the various atoms in the molecule. In a typical photomolecular event, a quantum in the ultraviolet is absorbed, most of its energy going to raise the molecule from its ground state to its first excited electron state; but some of that energy goes into vibrational-rotational form. When the molecule eventually decays back to its ground state, the vibrational-rotational energy is lost in small steps. It may, in the case of solids or liquids, be transferred to other molecules via collisions; or it may appear in the form of low-energy infrared photons which manifest themselves only as heat. The major drop from the excited electron level to the lowest electron level is accomplished by means of the emission of a single photon. This process is *fluorescence*—light emission with a wavelength some 10 or 20 percent longer than that of the exciting radiation (see app. F). It follows absorption by a time interval of 10^{-8} seconds or less. Fluorescence essentially "converts" ultraviolet to visible.[5] It is also an important scientific tool for the study of molecular levels.

Consider more closely the electron states of an organic molecule—for example, glucose, which has twenty-four nuclei surrounded by ninety-six electrons. Some of the electrons are called nonbinding; they are held in tight orbits around a single nucleus and have little effect on the molecule's chemistry. Other electrons, rather than orbiting a single nucleus, form what are called *covalent bonds,* locating themselves in a region where they can be "shared" by two adjacent atoms. These bonds consist of pairs of electrons, aligned in such a way that their intrinsic spins are in opposite directions. This behavior is dictated by the Pauli exclusion principle (see App. D), which asserts that two electrons cannot be in the same state; but the state of an electron is specified not only by its orbit but also by its spin. An electron's spin, according to the laws of quantum mechanics, can have one of the only two possible directions. And so, in a given electron orbit there can be two and only two electrons—their spins being opposite. Two such electrons become thus "paired" in the region between two atoms, sitting in the orbit of lowest energy and providing a particularly stable configuration. They form the "glue" that holds the atoms together.

In many cases atoms are held together by "double bonds," that is, *two* pairs of electrons shared by the atoms. The second pair, which forms what is called the *pi* bond, must have a different orbit from that of the first pair, called a *sigma* bond.[6] The electrons of the sigma bond provide more of the force that binds the atoms and are more tightly bound themselves in the molecule.

The pi electrons are less tightly bound, more easily excited, and are more important with regard to light absorption. Indeed it takes photons in the far-ultraviolet (wavelengths much shorter than that of violet light) to excite sigma electrons. Absorption of light in the visible and near-ultraviolet is largely by pi electrons.

The lowest excitation states of a molecule are those in which one of the pi electrons is excited to a higher energy state, that is, a different orbital configuration. When this happens the spin of the excited electron may be either unchanged or reversed. There are correspondingly two types of excited states, called *singlet* and *triplet* (fig. 8.5). In the singlet state the two pi electrons continue to have opposite spin; in the triplet state their spins are parallel.[7] (The exclusion principle is not violated in the triplet state, because the excited electron has a different orbit than the unexcited one.) Usually the energy of the singlet excited state is somewhat higher than that of the triplet state. Sometimes a second singlet excited state is observed at a still higher energy.

The triplet state is important in photochemistry and photobiology because it has the property of "trapping" a molecule, keeping it in that state for a relatively long time before it can decay to the ground state. A molecule reaches the triplet state in two stages. It is excited to the first singlet state by absorbing a quantum, and from this it falls into the triplet state by emitting one or several low-energy quanta. The triplet state then acts as a trap, because the laws of quantum mechanics forbid a transition by an isolated molecule directly to the ground state, by emission of a single photon. Various events (such as a collision between two molecules) eventually bring about the decay from the triplet state, but the times for these transitions are very long on an atomic scale, typically from about a thousandth of a second up to seconds, minutes, and hours.

Delayed emission from the triplet state produces the phenomenon known as *phosphorescence*. Certain materials, on exposure to light, begin to emit light of longer wavelength, and this light continues for a time after the stimulating light is turned off. Various sulfur compounds (for example, barium sulfide, the first known phosphor) are phosphorescent, and the color, the intensity, and the duration of the phosphorescence may be altered by placing traces of heavy metals in the materials. In any case, the wavelength of the phosphorescent light is independent of the wavelength of the stimulating light, since it is determined only by the energy of the triplet state.

An excited molecule has another possible course of action besides fluorescence and phosphorescence. It may undergo a chemical reaction, which would not be possible without the additional supply of energy. In many chemical reactions the energy of the products is higher than that of the reactants; an energy input

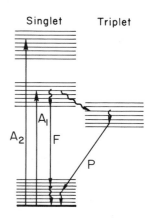

Singlet   Triplet

Fig. 8.5. Molecular transitions. Each molecular state is represented by a horizontal line; the vertical position is a measure of the energy of the state. The lowest group represents the electron in its lowest orbit; various vibrational and rotational substates are shown. Three excited electron states are shown; two are singlet states and one is a triplet state. Each of these has its own series of vibrational-rotational substates. A transition from one level to another is shown by an arrow. The set of transitions between two electron states forms a spectrum that is not a line but a band—a range of wavelengths. Transitions A_1 and A_2 are light absorption. The transition F is emission of fluorescent light. The energy emitted in fluorescence is less than that absorbed in the process A_1 that originally excited the fluorescence; some of that energy is lost in smaller infrared transitions (shown by the wavy lines) among vibrational-rotational substates. Phosphorescence is the transition P.

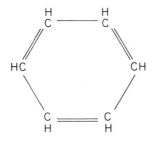

Fig. 8.6. The chemical structure of benzene. Six carbon atoms (C) are connected in the form of a hexagon by alternating single and double bonds. Each carbon also has one bond to a hydrogen atom (H). The rule in constructing organic compounds is that each carbon atom can have four bonds. This allows it to bond with atoms in a number of directions, so that complex molecules can be formed in a variety of geometrical shapes.

(sometimes in the form of heat) is necessary for the reaction to proceed. Other reactions produce an energy output but require a small starting energy (called the "activation" energy) to trigger the reaction. (The burning of a leaf releases energy but requires a small amount of activation energy—the match.) Light can supply either the energy input (as in photosynthesis) or the energy of activation (as in vision). We may picture a molecule with an excited electron as a structure with an extended probability cloud. It is more able to reach out to interact with other molecules and more likely to transfer that electron to another molecule or to share it in the form of a new covalent bond. Because an excited molecule sits in the triplet state for so long, that state is often the springboard for such photochemical processes.

But to use the light of the sun for these purposes nature must take special steps. Most organic compounds, as we have seen, absorb only in the ultraviolet. (They are either transparent or else reflect incident light diffusely and appear white.) To absorb that radiation which we call visible and which is abundant at the bottom of the atmosphere, special structures, called pigments, had to evolve. The characteristic of a pigment is a series of alternating single and double bonds among the carbon atoms of the molecule. The pair of electrons in the second bond (the pi electrons) are weakly bound, so that in a series of such double bonds these pi electrons become freed from their particular sites between two atoms and can wander across the carbon backbone of the molecule. Put another way, the probability distribution is no longer confined to the region of space between the two atoms but is spread widely, along the entire molecule. It is thus easier to excite these electrons, and so the absorption spectrum moves to longer wavelengths. (Why not use a molecule with all double bonds? Such compounds would be highly reactive and not suitable for a stable living system.)

A simple illustration of an organic structure with alternating single and double bonds is benzene, with six carbon atoms arranged in a hexagon (fig. 8.6). Benzene itself absorbs in the ultraviolet (the liquid is transparent), but larger molecules built from this hexagonal element constitute many of the common dyes.[8] Retinal, the pigment that mediates vision, has a chain of carbons built of such alternating bonds.

It would be misleading to conclude, however, that nature simply took note of what sort of photons reached the earth and then found a way to absorb them when needed. Rather, nature is severely constrained by the physical properties of the matter out of which protoplasm is formed. Life is built upon carbon, an element practically unique in its ability to form large linear and polygonal structures with the other common elements—hydrogen, oxygen, nitrogen, etc. (Certain other elements, silicon, for example, could in principle be used in the same way; but at earth's temperatures the compounds of silicon are not soluble in water.) An electron is an electron, with a unique mass and electric charge, and these parameters go a long way toward determining the energy levels of a molecule—hence the wavelengths that can be absorbed. Thus, given "life as we know it," the excitation of the molecules of life *must* be mediated by photons in the region from about 250 to 1,400 nm, regardless of what the sun puts forth. That there is an approximate match between the emission spectrum of the sun and the absorption spectrum of living things must not be regarded as inevitable; indeed, it is rather fortuitous.

Miscellaneous Responses to Light

Living things that see respond to light via the retina, the brain, and the rest of the nervous system. But there are many responses to light outside the nervous system, from the migration of granules of the pigment melanin to produce suntan, to the formation of vitamin D_3 in the skin when ultraviolet is absorbed. (Vitamin D_2, a related compound, is added to foods and is known to be effective in preventing rickets.) The squid changes color by the direct effect of light on muscle fibers. It has pigment in a small spherical sac, which, when tightened, spreads out into a disk and distributes the pigment's color over the squid's surface. Similarly, muscle contraction in the eyes of certain amphibia, fish, and the eel closes the iris in direct response to light, without the intervention of the nervous system. (In higher animals the iris is controlled neurally.)

Beneficial or necessary effects on human health from the natural spectrum of the sun are only now being discovered. One study suggests that moderate doses of ultraviolet help the body process minerals. It may be that artificial light, particularly fluorescent light designed to optimize vision, is not a satisfactory environment for all human functions.

Green plants bend toward the light *(phototropism)* and many microorganisms move toward or away from the light *(phototaxis)*. In bacteria and blue-green algae phototaxis operates through the same light-absorbing mechanism as photosynthesis—chlorophyll-like pigments which absorb in the red. These *prokaryotic* forms are the most primitive one-celled organisms; they lack a nucleus to contain their genetic material. Green algae are *eukaryotic,* higher one-celled organisms possessing a nucleus. Their phototactic response is to blue light, indicating absorption by a different pigment. Various pigments are involved in these processes, including the carotenoids, which are yellow pigments found also in higher plants.

Notice that when a plant bends toward the light, it is the part of the stem away from the light, presumably receiving less light, that must grow faster in order to make the stem bend. Phototropism must involve something more than merely differential growth rates, since one would expect the part of the stem receiving more light to grow faster. There is evidence that plant hormones, in response to light, migrate from one side of the stem to the other and stimulate growth on the dark side. (An exception is the mold *Phycomyces,* whose transparent stalk acts as a lens. Light from one side of the stalk is brought to a focus at the opposite side of the stalk, causing faster growth on that side and thus bending toward the light.)

Photoreactivation is the process whereby light may reverse many of the deleterious effects of ultraviolet, including mutation, inhibition of growth, and the killing of bacteria. Light of somewhat longer wavelength than the damaging ultraviolet is effective if introduced within a certain period of time (which may be several minutes or hours, or still longer if the cell is kept at low temperature). It acts in some cases by direct enzymatic repair of the original modification of the DNA molecule. Photoreactivation may have been an important device for the survival of organisms in the early history of biosphere.

Since sunlight varies periodically, both daily and seasonally, it is not surprising that these oscillations control many facets of life, and indeed such effects are seen at all levels of the phylogenetic scale. Research has shown, however, that a great many rhythmic biological functions, from eating and sleeping, to body temperature and hormone secretion, to luminescence in dinoflagellates (as we have seen), depend not only on the rising of the sun but also on an internal timing mechanism, the biological clock. Although it has been widely observed operating in microorganisms, fungi, and higher plants and animals (although not in bacteria), the clock is not well understood at the chemical level. The biological clock induces what is called a circadian rhythm in the function of organisms—that is, an approximately 24-hour cycle. The cycle determined internally may be shorter or longer than the environmental day by 1 or 2 hours, and the effect of light is to adapt the cycle to the exact 24-hour day and to fix when the cycle begins. This adaptation, called *entrainment,* can be demonstrated by keeping an organism in an artificial environment, with, say, 8 hours of light alternating with 8 hours of dark. The organism will follow the 16-hour day, carrying out its daytime functions (cell division, for example) during the light periods. But the circadian rhythm is innate. As soon as the artificial light is removed and the organism is kept not necessarily in a normal light pattern but, for example, in constant darkness, then the approximate 24-hour life cycle reestablishes itself. The biological clock of man can be upset by rapid transportation to a place of different

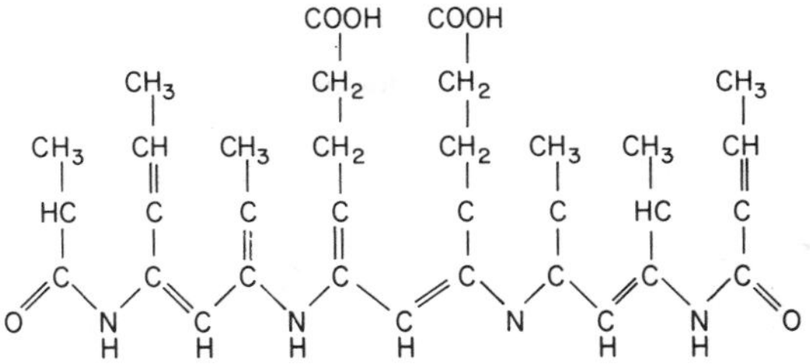

Fig. 8.7. The chemical structure of the pigment phytochrome, which mediates photoperiodism in plants. Shown here is the configuration P_r, sensitive to red light. A slight rearrangement converts it to the form P_{fr}, sensitive to far-red light. C, H, O, and N stand for carbon, hydrogen, oxygen, and nitrogen, respectively. Note many alternating single and double bonds, characteristic of a pigment.

longitude, but entrainment (along with psychological and cultural factors) brings him back into phase with the environment within a matter of days. Notice that entrainment is absolutely essential if the biological clock is to be effective at all in adaptation. If the circadian rhythm of an organism were, say, 25 hours, then within 12 days the organism would be completely out of phase with the sun.

Seasonal variation in the functions of animals and plants, controlled by the changing lengths of day and night, is called *photoperiodism*. Birds migrate in spring and fall; fish and reptiles mate in season; plants grow, flower, and fruit; trees become dormant in fall; and insects enter a diapause, (a state of suspended animation). The mechanism of these responses in animals is not well understood. But for plants, a pigment called *phytochrome* has been found to provide the mechanism for measuring the length of day and thus, in some way, to inform the plant of the time of year. The phytochrome molecule (fig. 8.7) responds to light, as retinal does, by changing its shape (although it does not separate from the protein to which it is attached, as in the eye). In one form (called P_r, for "red") phytochrome absorbs in the red, with an absorption maximum at 660 nm, and when it does, it is converted to the second form (called P_{fr}, for "far-red"). The second form absorbs in the far-red region, centered at 730 nm. During the day, photons at 660 nm are more abundant, and so P_r is converted to P_{fr}. In the absence of light, at night, P_{fr} reverts to the form P_r. As the seasons change the number of hours of daylight changes and the ratio of infrared to red intensity changes, and so there will be more or less P_{fr} in the plant. P_{fr} is responsible for initiating many plant activities, such as growth and flowering, while an accumulation of P_r signals a period of dormancy. Phytochrome is present in low concentrations in plants and so is not readily seen, but the isolated (and concentrated) compound is blue in the form P_r and gray in the form P_{fr}.

Photosynthesis

If we look, ultimately, at the process of life on the grand scale and how it intersects the events of the nonliving environment, we find

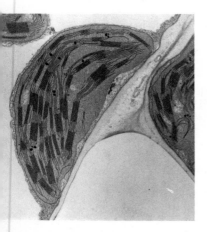

Fig. 8.8. Chloroplast in a leaf
of corn. Magnification 26,000.
(Micrograph by W. P. Wergin;
courtesy E. H. Newcomb)

there is one light-related activity that dwarfs all others in importance. This is *photosynthesis,* by means of which living things tap the energy of sunlight. Energy is needed for growth and repair, for mobility, for maintenance of temperature, and for many other vital processes. Although other sources of energy are available on earth (radioactivity, seismic energy, tidal energy deriving from gravitational forces), the source of energy relied upon by all living things is photosynthesis.

Organisms can be divided into two categories: those (mostly animals) that use chemical energy and those that use light energy. The former absorb chemicals of high energy content and, using enzymes to speed the reaction, convert them to chemicals of lower energy content, retaining the energy balance for their own functions. The chemicals absorbed are carbohydrates (and other molecules) deriving from other organisms and oxygen. The latter organisms are photosynthetic, taking in carbon dioxide and water (low-energy materials) and using the energy of light to form carbohydrates and oxygen, thereby creating themselves.[9] Photosynthesis thus drives the cyclic flow of matter (carbon, oxygen, and indirectly nitrogen and other elements) through the biosphere and the noncyclic flow of energy. Energy enters the system from the sun and leaves the system in the form of heat from decaying organic matter. Of the solar energy arriving at the earth, only a small fraction (0.05 percent) is used for photosynthesis. Of the energy thus used, a minute fraction goes to supply energy for the chemical consumers, most of it being used by the photosynthesizers for their own life activities. "Physiologically speaking, all the animals on land and in the sea, including man, are but a small brood of parasites living off the great body of the plant kingdom. If plants could express themselves, they would probably have the same low opinion of animals as we have of fleas and tapeworms."[10]

The action of animals in burning carbohydrates to produce carbon dioxide is called *respiration*. Photosynthesis and respiration establish an equilibrium in the gas content of air. That the present atmosphere contains 21 percent oxygen and 0.03 percent carbon dioxide (whereas the pre-biotic atmosphere contained no oxygen and probably a significant amount of carbon dioxide) is testament to the prevalence and efficiency of photosynthesis. On Mars and Venus, in contrast, the atmosphere is largely carbon dioxide and has little oxygen.

The major photosynthesizers are green plants on land and algae in the sea (the latter, in fact, account for more than half of the energy consumed). Land plants take carbon dioxide from the air through openings in the leaf and pass it into a chamber where it is dissolved in moisture and thus can enter a cell. Within the plant cell are small green structures called *chloroplasts* (fig. 8.8). The chloroplast contains a series of flattened membranous sacs, called *thylakoids* (fig. 8.9), sometimes stacked into layers of eight or ten, and surrounded by a viscous substance called the *stroma*. Both the stroma and the thylakoids play a part in photosynthesis, but the latter contain chlorophyll and the leaf's other pigments and support the light-related part of the process.

Among the photosynthetic algae the most primitive forms are the blue-green algae, also one of the most ubiquitous life-forms on earth. A prokaryotic organism with no chloroplasts, it nevertheless has thylakoid membranes, which capture sunlight in much the same way that green plants do, and it processes carbon dioxide dissolved in the waters. For millennia, beginning 3.5 billion years ago, blue-green algae dominated life in the sea, forming huge mats across the ocean bottom (fig. 8.10), and they maintained this dominance (all the while pouring oxygen into the air) until very recent times, some 600 million years ago, when higher forms—worms, mollusks, and other sea animals—began to prey upon them.

Still more primitive are the photosynthetic bacteria, filled with spherical photosynthetic units called chromatophores and able to function only in the absence of oxygen. In photosynthesis they use carbon dioxide (as their source of carbon) but not water, as we will see shortly. Because they never learned to tolerate environmental oxygen, they were condemned to remain in isolated regions and could not colonize the earth as did the blue-green algae.

The chemistry of photosynthesis in its most common form begins with the excitation of pigment molecules, which, in the excited state, catalyze a reaction in which water is split into hydrogen and oxygen. This first step is unique in biochemistry. In no other natural process is water split this way, so tightly bound is the

Fig. 8.9. Profiles of layers of thylakoids in a leaf of timothy grass. Magnification 100,000. (Micrograph by W. P. Wergin; courtesy E. H. Newcomb)

Fig. 8.10. Thick mats of photosynthetic blue-green algae growing on the floor of Lake Bonney in Antarctica. This may be very much like the form of life that dominated the seas for billions of years, until about 600 million years ago. (Photograph by Dale Andersen; courtesy George M. Simmons Jr.)

molecule. Energy to break its bonds is supplied by light. Oxygen is released to the air, and hydrogen atoms remain in the cell. Hydrogen nuclei (protons) go into water solution, and the electrons undergo an extensive series of reactions, being transferred from one compound to another. Light acts as a pump in this sequence, raising the electron to higher energy levels, until it is finally supplied to carbon dioxide (CO_2). Protons and CO_2 are then used to synthesize carbohydrates.

Carbohydrates make up a large class of organic compounds, including sugars, starch, and cellulose. A variety of carbohydrates are formed in photosynthesis, and more complex organic compounds are later built from these; but it will be simplest to describe formation of the simple sugar glucose, which has the chemical formula $C_6H_{12}O_6$. The equation describing the overall chemical reaction is

$$6CO_2 + 12H_2O \rightarrow C_6H_{12}O_6 + 6H_2O + 6O_2.$$

(O_2 is the oxygen molecule; CO_2 is carbon dioxide.) Six carbon dioxide molecules and twelve water molecules are needed to produce one sugar molecule; six oxygen molecules are released to the atmosphere. All twelve water molecules are needed for the process, although six are reconstituted at the end. For each oxygen molecule released, notice that four electrons must be transferred, one for each hydrogen atom split from water. It is thus expected that four quanta of light will be necessary, one to supply energy to each of these electrons.

In glucose and other carbohydrates the ratio of hydrogen atoms to oxygen atoms is 2:1, which suggested to nineteenth-century chemists that carbohydrates are made of carbon and water (as the name implies). In fact the oxygen in the carbohydrate comes from CO_2, not from H_2O; oxygen from the water molecule is released to the air. This fact was verified in experiments using the rare isotope oxygen 18 as a tracer. (Oxygen 18 is not radioactive; it can be followed in chemical reactions because it is heavier than the common form, oxygen 16.) Recall also that photosynthetic bacteria, which form carbohydrates, do so without using water at all. Rather they use other compounds, such as hydrogen sulfide (H_2S), liberating sulfur and obtaining electrons.

Let us turn briefly to the pigments. There are several varieties of chlorophyll, chlorophyll *a* being found most widely, in all photosynthetic organisms except bacteria (which have their own variant, bacteriochlorophyll). The chlorophyll molecule (fig. 8.11) has a ring constructed of carbon and nitrogen atoms, held together with alternating single and double bonds, with a magnesium atom at the ring's center. Attached to the ring is a long chain of carbon atoms, which is probably not involved in light absorption but aids in binding the molecule to the thylakoid membrane.

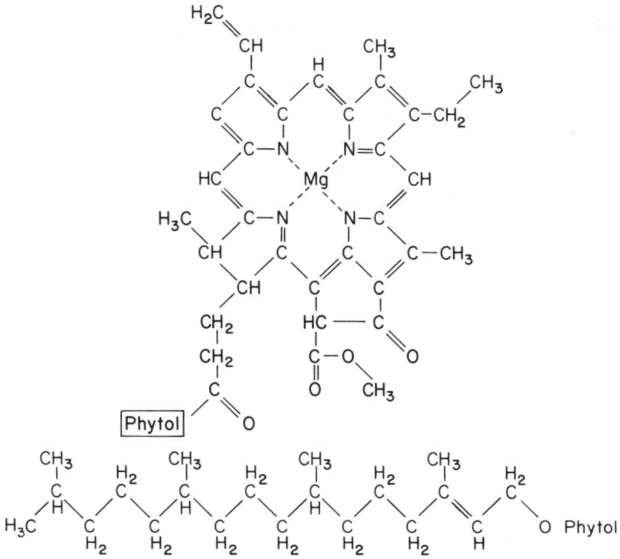

Fig. 8.11. The molecule of chlorophyll *a*. The light-absorbing portion contains alternating single and double bonds, forming a series of rings surrounding magnesium (Mg). The phytol is a side chain, which has only a structural (rather than a light-absorbing) function.

The absorption spectrum of chlorophyll *a* shows peaks in both the blue and the red, so that only the intermediate wavelengths, predominantly the green, are transmitted. The red absorption corresponds to exciting the molecule to its first excited singlet state; the blue to the second excited singlet state. (It was held for some time that the triplet state was important in the photochemistry of photosynthesis, but evidence indicates that it is not.) Chlorophyll in the leaf exhibits a weak fluorescence when illuminated, while chlorophyll in solution is strongly fluorescent. In other words, in the leaf the excited electron has a small probability of falling down to the ground state and a high probability of proceeding along a chemical pathway (that is, of being transferred from molecule to molecule). Removed from the living organism, the electron finds the chemical route no longer open and so can only decay. The fluorescent light is red, even if the illumination is blue, indicating that excitation to the second excited state is only transitory and that the electron quickly falls to the first excited state (with loss of heat) and from there falls to the ground state.

Other pigments include the *carotenoids,* found in higher plants, bacteria, and algae. There are several varieties of carotenoids; most are colored yellow (sometimes orange or red), and they absorb in the blue to blue-green.[11] The molecule is a long carbon chain, with alternating single and double bonds; it is related to retinal (fig. 8.12). Illuminating the pigment system with light wavelengths capable of being absorbed by a carotenoid and not by chlorophyll still produces a red fluorescence characteristic of chlorophyll. Thus the accessory pigments are able to transfer their

171

energy to chlorophyll, allowing the organism to use light over a wider range of wavelengths to drive photosynthesis. There is also evidence that the carotenoids can protect chlorophyll from destruction at high light levels.

We picture the chemistry of photosynthesis as made up of two stages: (1) the splitting of water to release electrons and (2) the use of those electrons at a higher energy, together with CO_2, to build carbohydrates. In an important early experiment, R. H. Hill demonstrated the first stage without the second. He illuminated powdered leaves in water and found that they released some oxygen but produced no carbohydrate. Furthermore, by adding certain iron salts that are more eager receivers of electrons than CO_2, he increased the production of oxygen. The dead leaf lacks the enzymes to build carbohydrates, but, given sufficient impetus, the water-splitting reaction still proceeds.

Light acts as a pump, pushing electrons "uphill," that is, to higher energy. But there are many intermediate steps in the electron transfer process (not all of which are understood), and there is compelling evidence that the pumping effect of light occurs in two steps. Certain of the intermediate compounds (molecules that receive and then donate electrons) are known, and it is also known that their energy is intermediate between that of water and that of carbohydrate. Thus light must first raise an electron to the intermediate state and then raise it again to the final state (fig. 8.13). (Various chemical reactions—electron transfers—go on before and after each absorption of light, but in these the electron falls downhill, some energy being lost to the medium as heat.)

Further evidence comes from measuring the quantity of light energy necessary to evolve a given volume of oxygen. This may be expressed in terms of the "quantum requirements," the number of quanta necessary to evolve one molecule of oxygen. This number turns out to be very close to eight, showing that each of the four electrons that must be transferred needs two quanta.

A number of ingenious experiments have shown, in fact, that there are two independently functioning photosystems, each with a slightly different absorption spectrum (although both use predominantly chlorophyll *a*). The first has its absorption maximum at 680 nm; it splits the water molecule, frees its electron, and provides the first electron boost. The second, with absorption

Fig. 8.12. The molecule of β-carotene, one of the carotenoid pigments. It is related to vitamin A and is also similar to the retinal molecule (see fig. 2.26).

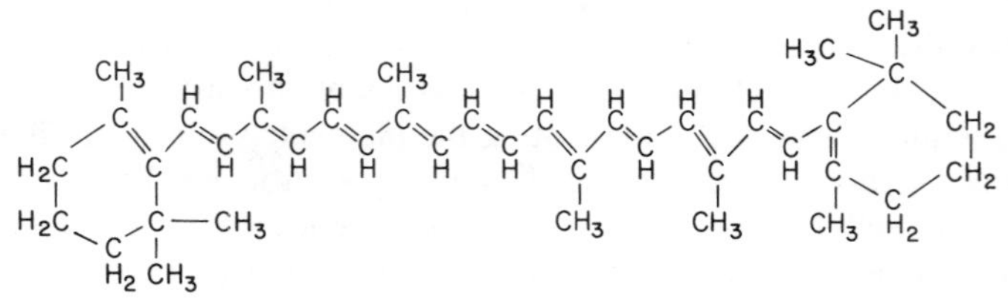

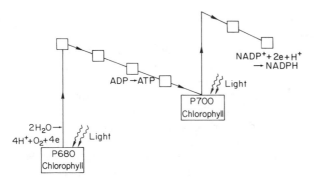

2H$_2$O →
4H$^+$+O$_2$+4e

P680
Chlorophyll

ADP → ATP

NADP$^+$+ 2e + H$^+$
→ NADPH

Light

P700
Chlorophyll

Light

Fig. 8.13. Schematic view of the light reactions in photosynthesis. Vertical position represents the electron energy. *P680* and *P700* are the reaction centers for the photosystems that have absorption peaks at 680 nm and 700 nm, respectively. In *P680* water is broken apart with the aid of energy from light; electrons are boosted to a higher energy state. The small squares represent a complex of molecules that receive and/or transmit electrons. In these next reactions the electron loses energy and ATP is manufactured. Then *P700*, also absorbing photons, provides the second energy boost. The final electron acceptor is NADP$^+$.

maximum at 700 nm, provides the second electron boost. In photosynthesis experiments, if the wavelength of light is taken beyond 680 nm, to 690 or 700 nm, the reaction, as measured by production of oxygen, drops off rapidly. Put another way, the quantum requirement becomes very large. But suppose one adds to the chloroplast what is called an artificial donor, that is, a substance that donates electrons much more readily than water. Ascorbate is one example. Now, if the wavelength of light is 690 or 700 nm, the quantum requirement *drops* to four. The artificial donor has bypassed the first photosystem, which is inoperative. The first boost is not needed; only the second, which is most effective at 700 nm, is used. (The electron in ascorbate, already at a higher energy than the electron in water, needs only the second boost.) On the other hand, at shorter wavelengths, where both systems are absorbing, the chloroplast goes back to using electrons from water, and the quantum requirement is eight.

Experiments in the 1930s showed that a minimum of about 2,400 chlorophyll molecules is needed to produce one oxygen molecule. Since each oxygen requires eight photons, it becomes useful to think of a group of about 300 chlorophyll molecules as making up what is called the photosynthetic unit, a minimal quantum-absorbing unit within the thylakoid membrane. (The size of the photosynthetic unit is not fixed, however. It varies, among other things, with the intensity of light under which the plant is grown.) It is presumed that there is one "reaction center" in one of these complexes, a site where the chemistry of photosynthesis begins, and that the other molecules in the unit act as antennae; they form a light-receiving system designed to transfer energy through the unit until it reaches the reaction center. (Or if it does not, it is emitted eventually as fluorescence.) Evidence for this view comes from chlorophyll fluorescence studies under polarized light. If the same molecule that absorbed a photon later emitted it, then the fluorescent light would also be polarized. In the experiment fluorescent light is found not to be polarized, indicating that transfer of the energy of excitation had occurred before light was emitted.

If a photosynthetic organism receives a very short flash of light (say, 0.0001 second), photosynthesis, as measured by oxygen production, proceeds for some hundredths of a second. Thus there is both a *dark process* (a series of reactions) that can continue without light and is relatively slow, and a *light process* that provides ingredients necessary for the dark reactions. The light reactions, as we have been discussing, provide electrons at a high level of energy, and these are finally fed into carbon dioxide to build carbohydrate. The final donor, a molecule called nicotinamide adenine dinucleotide phosphate (abbreviated $NADP^+$), receives two electrons from the electron transport system of the light process and draws one proton from the aqueous medium to form NADPH, which enters the realm of the dark process carrying these electrons. The light reactions also synthesize ATP using some of the excess energy absorbed by the reaction center of the photosystem whose absorption peak is at 680 nm. These light reactions supply this molecule as an energy source to the dark reactions.

The dark process is the best understood phase of photosynthesis, partly because its processes are relatively slow and also because they take place not in the membrane but in the stroma of the chloroplast, where all the chemicals involved are in solution. In the laboratory of Melvin Calvin, leaves and algae were studied in an environment of radioactive carbon dioxide—that is, CO_2 formed with the radioactive isotope carbon 14. Since isotopes of an element behave in the same way in chemical reactions, the sequence of events in the dark reactions can be followed by exposing organisms to radioactive CO_2 for varying amounts of time and then monitoring the formation of complex compounds built of carbon 14.[12] Within seconds algae begin to produce a variety of radioactive compounds, and soon photosynthesis produces not only sugars and starch but other substances not classified among carbohydrates: fatty acids, fats, and amino acids. The first stable intermediate compound is *phosphoglyceric acid,* a combination of phosphoric acid (an inorganic acid related to the reactions of ATP) and glyceric acid. The formula of the latter is $C_3H_6O_4$, placing it intermediate between CO_2 and glucose with regard to the degree to which carbon atoms have been assembled.

Many of the steps in the dark process are reminiscent of similar reactions in the process known as *glycolysis,* a process that also makes use of phosphoglyceric acid and is widely involved in energy *consumption* in the biological world. That is, a series of similar reactions is used in the breakdown of sugars to release CO_2 and to package energy in the form of ATP. Indeed it has been suggested that glycolysis formed the (or a) biochemical basis of life before photosynthesis evolved, when organisms lived off earth's nonbiological chemicals, and that photosynthesis subsequently evolved by modifying the glycolysis pathway and grafting onto it new structures able to capture the energy of the sun.

Finally there is the matter of evolution and the presence of *deoxyribonucleic acid*. Strands of DNA are found within the chloroplast; there is not enough DNA to allow the chloroplast to replicate itself, but there is enough to encode for a few hundred proteins. Cultured outside the cell, chloroplasts can carry out photosynthesis and even manufacture some proteins, but only for a very short time. Many biologists believe that the chloroplast is a remnant of an independent prokaryotic cell, which, in the distant past, was engulfed by a nonphotosynthetic cell and established a symbiotic relation with it. The captured cell, possessing the secret of photosynthesis, may have been an ancestor of present-day blue-green algae. More precisely, a widely held modern theory is that the eukaryotic plant cell evolved from an association of prokaryotic cells living within a host eukaryotic cell, each cell performing a specialized function, and that the evolution of this association of cells led eventually to the highly efficient photosynthetic organisms of today, their parasites in the animal kingdom, and the profound effects wrought upon the entire planet.

High Technology

9

When man and woman learned to control fire they learned to control light. The purposeful manipulation of light probably dates from earliest human civilization; and advances in the control of light have closely paralleled the growth of civilization, from the telescope of Galileo to the microscope of the pathology laboratory, from Edison's electric bulb to today's array of lasers. Control of light means two things: manipulation of existing light sources and creation of new sources. In the manipulation and use of our richest source, the sun, we have been quite backward.

Plants of course, as we have seen, are not at all backward. Absorbing only a small fraction of incident sunlight, they manage to use it for all their needs, and to provide for our biological needs as well. Primitive human beings needed 2,000 calories per day—their ration of food. When they tamed fire their energy consumption went up to 5,000 calories; and when they began to use animals extensively in agriculture and to harness the power of falling water in industry, they consumed 26,000 calories. These forms of energy are all derived ultimately from the light of the sun through processes that are, primarily, the inventions of nature. In our modern technological society we require about 230,000 calories for each individual and so have had to resort increasingly to our limited supplies of fossil fuels and uranium.[1] The energy of coal, oil, and natural gas also derives from the sun; it is the partially degraded energy once assembled by green plants. But since we use it at a rate millions of times faster than it is being collected on earth, it must be regarded as a fixed and limited source, which will, before too long, be depleted. Thus attention has been focused on capturing sunlight as it arrives at the earth.

Solar Energy

The problem basically is one of transformation and storage: we cannot grasp a photon or keep it in a box. In a rooftop solar collector we transform the energy of sunlight into heat by raising the temperature of some transfer fluid, such as water. Heated water can be used directly or can transfer its heat to the air and walls of a room. The problem with direct solar-thermal conversion is that

Fig. 9.1. *Euphorbia lathyris* growing in Santa Ana, California. This plant has been suggested as an "energy crop." (Lawrence Berkeley Laboratory, University of California; photograph by Gene Elle Calvin; courtesy Melvin Calvin)

sunlight is not always available, varying as it does with season, time of day, and weather. Heat may be stored for a time, but soon dissipates, and is not ready for your bath. One way to collect and employ stored solar energy is to further parasitize the plant kingdom. Some scientists have proposed the planting of "energy farms," where crops are grown to be burned. Sugar cane is very efficient in converting sunlight to chemical energy; sweet sorghum and corn, which grow more readily in the United States, are also effective photosynthesizers. Such materials, renewed each year, could be burned instead of coal and oil in power plants to generate electricity. They can also be fermented to produce alcohol, a superior fuel because the carbon in it is less oxidized.[2] Still better as a fuel are hydrocarbons, organic compounds containing no oxygen. Fossil fuels are of this type, but some plants carry photosynthesis an extra step (or series of steps) and produce hydrocarbons. The Hevea rubber tree, grown in Malaysia and Indonesia, is such a plant. It produces a latex that, when dehydrated, leaves an elastic solid, rubber. Related plants of the genus *Euphorbia,* in a similar way, produce liquid oils that may be economical as fuels (fig. 9.1). Research indicates that these plants can be grown in the United States, and, unlike sugar and corn, they do well in semi-arid regions. This is advantageous, since these "energy crops" would have to compete with food crops for arable land. Indeed, many experts doubt the practicality of energy farms because of the scarcity of land and water. Other more speculative approaches to solar energy are based on imitating nature rather than stealing from her. For example, since photosynthesis is a photochemical process involving the transfer of electrons at the thylakoid membrane, one might seek an artificial membrane that would, in a similar way, produce an electric current.

Less speculative, but still in the future, is large-scale electricity generation based on semiconductors—the *photovoltaic* (or solar) *cell*. The solar cell of course is already with us, having been used successfully as a source of power in spaceflight. It offers the advantages of fuel-less generation of electricity and easy deployment in either large or small (single home) installations and is practically maintenance-free (since there are no moving parts). At present the cost of electricity from photovoltaics is high, making it noncompetitive with conventional sources. When and whether it will be competitive depends on a variety of factors both technical (how cheap will solar cells be in large-scale production) and economic (when will industry embark on large-scale production), but the scientific nature of these devices is well-understood. They are based on the properties of the *semiconductor,* the material that lies at the heart of the current revolution in communications and that will play many roles in the new high technology of light.

A semiconductor (see app. G) is a crystalline solid in which the atomic electrons are normally not available to carry an electric current. They can be made available by flooding the solid with energy, either in the form of heat (raising it to high temperature) or light. When a photon is absorbed in the solid, one of the electrons may be raised to a state in which it is free to move through the lattice. The trick is to provide an environment in which the electron will be made to flow through the solid before it has a chance to decay (with reemission of a photon) back into the state from which it was excited. Such an environment exists in *p-type* and *n-type* semiconductors, artificially produced materials with opposite electrical properties.

Starting with a semiconductor such as silicon, we produce an *n*-type semiconductor by adding to it, in a controlled way, small quantities of an element such as phosphorus. This process of adding "impurities" is called *doping*. Phosphorus has five outer electrons, whereas silicon has four, so that the effect is to supply the solid with a small number of extra negative charges, electrons, which are relatively free to move about in the lattice. A *p*-type semiconductor is made by adding to silicon impurity atoms that have three outer electrons. Thus there are "holes," places where electrons normally would be but are not, and these entities behave like positive charges, similarly free in the lattice. A junction between these two types of semiconductors is therefore electrically asymmetric; we may think of it as in a state of tension, like the stretched string of a bow. This tension is, to be precise, an electric field, and when electrons are made available, they respond to the field by flowing through the crystal, from the *p*-side of the junction to the *n*-side. This flow is the electric current generated by light.

Doped semiconductors are widely used, but their manufac-

Fig. 9.2. Phase relations. A quantity such as an electric field (plotted on the vertical axis) oscillates with time (the horizontal axis). The figure shows several oscillators with different phase relations to each other. Oscillator (*c*) is "out of phase" with oscillator (*a*): whenever (*a*) has a crest, (*c*) has a trough. Oscillator (*e*) is "in phase" with (*a*): their crests and troughs come together. If we represent one complete cycle by 360 degrees, we say that (*c*) and (*a*) differ in phase by 180 degrees. But (*e*) and (*a*) differ in phase by 360 degrees, which is equivalent to 0 degrees. A closer look shows that (*b*) differs from (*a*) by 90 degrees (one-quarter of a cycle), and (*d*) differs from (*a*) by 270 degrees (three-quarters of a cycle). As long as the phase relation between two oscillators is *fixed*, they are coherent. But if the phase relation varies randomly—sometimes in phase, sometimes out of phase—the sources are incoherent, like lights from two different lamps falling on a surface.

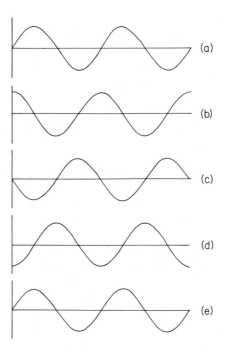

ture is expensive. Natural minerals contain unwanted impurities at much higher concentration levels than those eventually to be implanted by doping. The technological problem is to produce an extremely pure silicon crystal, to add controlled quantities of *p*-type and *n*-type impurities, and then to lay a very thin layer of *n*-type semiconductor upon a substrate of *p*-type (thin in order that light can penetrate to the junction). Depending on how these technologies develop (and how they compare with competing energy strategies), the twenty-first century may see solar cells not only on the rooftops but perhaps also in vast arrays in space, collecting sunlight and beaming energy in the form of microwaves down to earth.

The Laser

The doped semiconductor has been with us for several decades. Along with two other recent inventions, the light fiber and the laser, it will form the foundation of the coming revolutionary technology of communications and energy. Behind these systems lies the concept of *coherence*—which we have discussed only briefly—as it applies to light.

Ordinary sources of light—an incandescent bulb, a reflecting wall, or the sun—are spread out across our field of vision. The eye collects light from different parts of the source (i.e., light emitted by different atoms). There is no synchronization between emissions from any two parts of the source, so that, although the two wave trains may be very similar (in color and intensity, for example), they differ in phase in a random way (fig. 9.2). This difference is not apparent in observations such as vision and photography, but it can be seen in interference phenomena, since they depend on two light beams being in or out of phase. Consider, for example, the two-slit experiment of Thomas Young, the early demonstration of the wave nature of light. If Young had looked for interference between light beams from two small bulbs, he would have failed. The experiment works because the two light sources, which are the two slits, are actually created by a single *small* light source behind the slits. Suppose the single source is equidistant from the two slits. Then every time the crest of a wave passes through slit 1, the crest of a wave is also passing through slit 2. It follows that at certain points along the receiving screen the two waves will interfere constructively; at nearby points they interfere destructively, creating thereby the characteristic fringes of an interference pattern. Light beams from the two slits in this instance are said to be coherent; light from two different bulbs, or from two parts of an extended bulb, is incoherent (see fig. 9.3).

We create coherent sources at the expense of intensity. The source behind the two slits must be small, and so a very limited quantity of light is available for the interference experiment. Here lies the unique contribution of the laser, for it is an intense source of light, produced in such a way that all light across the breadth of the beam is coherent. There are four characteristics that distinguish laser light from conventional light: coherence, monochromaticity (which is in fact related to temporal coherence[3]), directionality, and intensity. That laser light is emitted in an extremely narrow range of directions is a consequence of the geometrical arrangement of the laser tube. That laser light is coherent is a consequence of a fundamental property of quantum mechanics.

The property is called *stimulated emission,* part of the expression "light amplification by stimulated emission of radiation," out of which the acronym "laser" is made. The idea goes back to a 1917 paper by Albert Einstein, who had continued to reflect upon the quantum and the transition between atomic energy levels. Two types of transitions had become understood through Bohr's theory of the atom. An atom can be excited from a low level, with energy E_1, to a higher level with energy E_2, by absorbing a quantum, whose frequency, ν, was given by $h\nu = E_2 - E_1$; and an atom can *spontaneously* emit radiation of the same frequency by falling from the upper level to the lower. Einstein presented an indirect but powerful argument for a third type of transition, stimulated emission, in which an atom in the upper state is *forced* to emit radiation, and to decay, by the presence in its vicinity of a photon of this same frequency (see app. H). The original photon still exists after the stimulated emission, and both photons may go on to stimulate emission in other atoms. Provided there are enough atoms in the upper state, there can be a kind of chain reaction, in which photons of the appropriate frequency produce more and more photons like themselves.

To be more exact, in the presence of this radiation, all three processes, absorption, spontaneous emission, and stimulated emission occur simultaneously. Absorption requires that there be atoms in the lower energy state; emission requires that there be atoms in the upper state. It is therefore possible for emission to "win" and drastically increase the number of photons, if there are more atoms in the upper state than in the lower. Here lies the diffi-

Fig. 9.3. In this experiment laser light falls on a square array of small holes. In one case the light is coherent over the whole array, and the diffraction pattern on the left is produced. If the light is coherent only over a small area (smaller than the spacing between the holes), then the pattern on the right is produced. (Courtesy Brian J. Thompson)

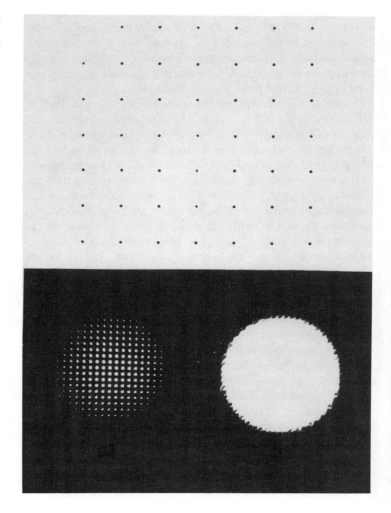

culty in making a laser. For while stimulated emission is a natural process in any atomic system, it is unnatural to have more atoms in a higher energy state than in a lower energy state. The natural state of affairs is for most atoms to be in the ground state, a few in the first excited state, and fewer and fewer in the subsequent higher states. If the temperature is increased, we find more atoms in higher states, but the normal order is always that the "population" is smaller and smaller as the energy level increases. Thus the laser, which produces large quantities of photons by stimulated emission, requires the existence of what is called a *population inversion*. (The normal order of population exists for a system at equilibrium; the laser is not at equilibrium.)

Various designs for creating a population inversion have led to a wide variety of types of lasers, with different materials and different properties, including a range of wavelengths from the ultraviolet through the visible and infrared. As one example, consider the helium-neon laser (fig. 9.4). Laser light is emitted from a transition between two excited neon levels (not a transition to the ground state). The key to laser action lies in the lifetimes of these levels, the time it takes before they decay. The upper level, call it *B*, has a relatively long lifetime, and so atoms tend to accumulate in it. The lower level, *A*, has a short lifetime, so that any atoms that fall down from *B* to *A* are quickly dispersed downward, to the ground state or to another level below *A*. Population does not accumulate in *A* and so an inversion occurs. There must be a system for "pumping" atoms up to the higher state, *B*, and the ultimate source of this energy in the helium-neon laser is an electric discharge through the gas. Helium is used as an intermediary here and in a number of types of lasers because it has several long-lived levels in which energy is trapped and stored. The upper level (*B*) of neon happens to be very close to one of these helium levels, and so energy is easily transferred from helium to neon via collisions. The continuous flow of electric current generates a continuous beam of red or infrared radiation from the neon transitions.

Another type of laser is made with a mixture of helium and various metal vapors—for example, the helium-selenium laser. It emits at many points along the visible part of the spectrum (and in the ultraviolet) because selenium has many pairs of levels in which a population inversion may be induced. Carbon dioxide and nitrogen make a powerful and unusually efficient laser, based on transitions between vibrational-rotational levels of the CO_2 molecule. The first laser, built by T. H. Maiman at the Hughes Aircraft Company, made use of an artificially grown ruby crystal pumped by an extremely intense flash of light. The excitation in this case was sufficient to create an inversion between an excited state and the ground state (a very unusual arrangement for a laser)

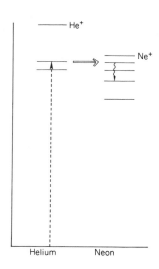

Fig. 9.4. The helium-neon laser. The figure shows some of the energy levels of helium and neon. The bottom line is the ground state. An electric discharge in the gas tube excites (*dashed arrow*) helium to a long-lived level. This energy is transferred (*horizontal arrow*) to the neon atom in a collision. Neon makes a transition (*wiggly arrow*) to one of its lower excited states. This neon transition becomes the stimulated emission of the laser. The levels He^+ and Ne^+ are the energies of ionized helium and neon, that is, the atoms with one electron removed.

and an intense narrow beam of red light for a thousandth of a second. Some lasers act in this mode, with brief highly concentrated bursts of energy. Others emit continuously.

An important laser property for some applications is *tunability,* the possibility of choosing its wavelength and varying it over some range at will. An example is the dye laser, in which the lasing medium consists of pigment-like molecules dissolved in water or some other solvent. The transition is between an excited vibrational level of the first excited electron state and an excited vibrational level of the ground state. Since there are many closely spaced vibrational levels (with rotational sublevels also), the possible laser wavelengths form an almost continuous band. The designer chooses the wavelength by means of a device, like a diffraction grating, that selects only one wavelength for reinforcement in the dye solution. (The key here is that those vibrational levels that do not become reinforced, rather than emitting light by spontaneous emission, are more likely to fall in small steps through the vibrational levels and down to the one that has been selected for lasing. Thus all the excited molecules participate in laser action, and the intensity remains high.)

Laser light is intense because of the multiplicative effect of stimulated emission; each photon, traveling through the material, produces new photons and is not itself destroyed. Moreover, the material is enclosed between two mirrors (for example, a gas laser is simply a tube containing the gas, with parallel mirrors at the ends), so that photons continually bounce back and forth, stimulating the creation of new photons many times. One of the mirrors is not completely reflecting; it transmits some fraction of incident light, so that the beam emerges from the tube. The wavelength of the stimulated photon is precisely the same as that of the photon that produced it (like *cloning* the first photon), so that the beam reaches extremely high monochromaticity, one of the most striking features of laser light. If the light is spread out into a spectrum, we observe an extremely narrow "spectral line."

In comparison, an ordinary gas discharge tube also produces a line spectrum (and the light of one line is monochromatic), but to get an intense beam we would have to step up the electric current, and that would raise the temperature. High temperature in turn makes the spectral lines broader because of the Doppler effect: random thermal motions of the gas atoms change the wavelengths of emitted light. Laser light, on the other hand, is made more intense simply by allowing more stimulated emission, without raising the temperature of the gas.

The remarkable directionality of laser light is a consequence of the geometry of the system, the two parallel mirrors at the ends of the tube. Laser action depends on many many reflections back and forth along the tube, and any photons that travel in a direction *not* along the axis are eventually lost through the tube's sides. The resulting beam is thus a narrow pencil of light that, unlike a flashlight beam, diverges very little after it leaves the source. The narrowness of the beam is, in another sense, the reason for its intensity. A typical gas laser may be only a few watts in power, but its energy is so concentrated into a small spatial area that the laser beam in your eye can cause serious damage; the light of a 100 watt bulb is diffuse, and only a small fraction of its energy enters the eye. (The danger of the laser is not so much that you will point it into someone's eye, but that the beam will inadvertently be reflected in the wrong direction by some mirror-like surface.)

Most important perhaps, stimulated emission produces coherent light. The time of creation of the stimulated photon is determined precisely by the time of arrival of the first photon, and so the two waves proceed down the tube "in step," with crests together and troughs together. Since a coherent beam is capable of producing interference and diffraction effects characteristic of light waves, the advent of lasers has opened up an entire new technology based on these effects.

Uses of the Laser

We first saw diffraction as the faint light and dark bands along the edge of a shadow; or the faint light and dark circles surrounding a circle of light that has passed through a small hole. Suppose we look carefully at the light that falls on a screen after passing through a circular hole in an opaque barrier. As the hole is made smaller, the circle of light on the screen becomes smaller, and the diffraction fringes change. But beyond a certain point (and this depends on the light's wavelength), as the hole decreases in size, the diffraction pattern expands. Through the larger holes light travels approximately in a straight line, and the circle on the wall approximates the hole itself. But light emerging from a very small hole spreads out as a series of wave fronts, the more so as the hole becomes smaller. Figure 9.5 shows the progression of diffraction patterns for a series of circular apertures of different sizes.

The inverse relationship between the size of a small object and the size of its diffraction pattern allows us to use diffraction as a tool for "seeing" the object, measuring its size, for example, or detecting small errors in its shape. The idea goes back as far as Thomas Young, who used it to measure the diameter of wool fiber. It becomes a practical device for use in industry because the laser makes the diffraction pattern intense and easily detectable. Fibers, small particulates in polluted air, droplets in aerosols, needle points, and blood cells are examples of systems that have been analyzed in this way (fig. 9.6). The shape of the extended light pattern may be determined in an automated on-line system in which diffracted light is "read" by an array of semiconductor photocells. The objects being measured may be moving on an assembly line, and the current produced in the photocell can be used to activate a robot arm to discard defective products.

Other applications of coherent optics make use of interference patterns created by the index of refraction of a transparent object. Normally an image is a representation of the way an object absorbs light. For example, in a photographic negative the dark places absorb light and the light places transmit light, and when we make a positive print we are recording the *intensity* pattern of light passing through the negative. But a transparent object also changes the *phase* of transmitted light if the speed of light is reduced as it passes through. One example of such an object is a living cell, which retards the light beam in a transmission microscope. Another application is a device (a *schlieren system*) used to study hot gases, where the index of refraction is a measure of the gas temperature. Such phase effects are not detected by ordi-

Fig. 9.5. Diffraction patterns produced by circular apertures of various diameters. The *relative* sizes of the apertures are shown on this page. (The largest is 2.18 millimeters.) The corresponding diffraction patterns are on the right-facing page. For the larger apertures the lit part of the pattern is approximately the size of the aperture and decreases as the aperture decreases. Notice the dark spot in the center of some of these examples. For smaller apertures the diffraction pattern increases as the aperture decreases. (Courtesy Brian J. Thompson)

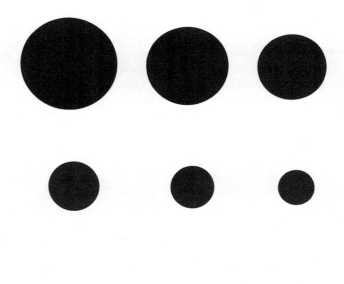

nary photographic methods, since film is sensitive only to the light intensity, not its phase. Rather the transmitted beam must be combined with a second beam that has not passed through the object. The interference pattern created between the two beams does depend on the phase, and this pattern is then recorded on film.

In the short time since the building of the first laser in 1960 the unique characteristics of laser light have found application in so many areas of modern technology that it is difficult to survey them all, no less to predict future developments. One of the earliest medical applications is the welding of a detached retina. The laser beam is actually focused by the lens and cornea of the eye, so that its energy falls on a very small area of the retina, attaching it to the layer behind, with little damage to surrounding cells. Moreover, energy can be delivered in a pulse lasting less than a thousandth of a second, so that the problem of head and eye movement during the operation is minimized.

Energy absorbed from a finely directed laser beam can cut

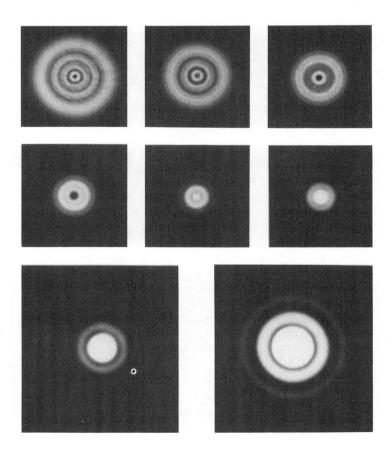

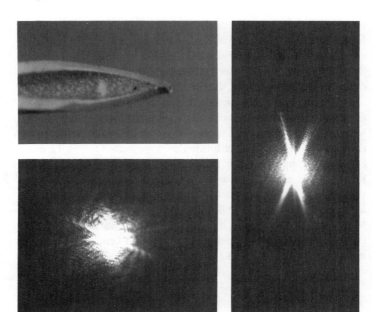

Fig. 9.6. The magnified photograph on the upper left is of a defective hypodermic needle; it has a small hook at the tip. On the lower left is shown the diffraction pattern of that needle. The figure on the right is the diffraction pattern of a perfect hypodermic needle. (Photograph reprinted, by permission, from Nicholas George, *Automatic Pattern Recognition* [1986], 43.)

tissue as well as weld two adjacent tissues. The beam automatically seals off smaller blood vessels as it cuts, reducing the time-consuming procedure of tying vessels and lessening the loss of blood. Risk of infection is greatly reduced; no germs ride on a photon. And the cut of a laser heals without scar tissue. Lasers have been used to seal blood vessels in cases of gastrointestinal bleeding and in cases where excess blood vessels develop in the eye as a consequence of diabetes (a measure found effective in preventing blindness). The laser beam can cut and reattach the tiny fallopian tube, which may be blocked as a result of various conditions, preventing conception. Other gynecological applications include a new effective treatment for excessive menstrual bleeding, based on vaporizing the inner layer of the uterus, a procedure much less radical than hysterectromy. In gynecology as well as in other fields the laser is being used to destroy cancerous or precancerous growths, allowing doctors to avoid removing entire organs.

The extreme directionality of the beam has made the laser an important tool in areas of research from the moon to the interior of a cell. The Apollo 11 astronauts placed a series of reflectors on the moon's surface to reflect laser signals sent from earth. Timing of the return signal (about 2.5 seconds round-trip) measures the distance to the moon with an experimental uncertainty of less than 1 foot. Careful measurements of this distance over time allow scientists to determine the lunar orbit precisely and to observe irregularities of the moon's motion about its center. These irregularities provide information about the distribution of mass in the moon's interior and hence about its geological history. Measurements from two points on the earth's surface allow geologists to determine the rate of drift of the continents. Laser reflectors placed in the crater of Mount St. Helens help establish the rate of growth of the lava dome. And at the opposite extreme, biologists can focus a laser beam through the optical system of a microscope to a spot smaller than a cell; this technique is used for internal cell surgery. Biologists can selectively destroy certain of the cell's subunits, such as the mitochondria, the chromosome, or parts of the chromosome, and thus observe effects on the cell's function.

The fine point of a focused laser beam can be used to clean art treasures, to trace handwritten messages or designs on a cathode-ray screen, to levitate microscopic liquid drops, and to store digital information on new ultra-high-density disks. Here the beam creates a sequence of microscopic pits, or bubbles, which embodies information in a binary code (the presence of a bubble represents the digit *1* and its absence represents *0*). The record can later be read by illumination, since the bubble is reflective. Ten billion bits of information can be stored on a disk 1 foot in diameter.

New laser techniques are being used for rapid analytical chemistry. The beam ionizes or excites atoms and molecules, and scientists observe characteristic fluorescence or spectral emission lines, fingerprinting the atom or molecule. Devices along these lines have been developed for detecting pollutants in air, for studying intermediate processes in the burning of fuels, and, by the military, for detecting the use of chemical weapons. There are also potentially important chemical applications based on the great spectral purity of laser light. Chemical reactions begin with the breaking of specific chemical bonds, which requires the input of a precise quantity of energy. Traditional chemistry supplies this energy inefficiently by adding heat, whereas a laser can be tuned to just the energy needed for the particular reaction.

A similar application arises from the need for separating isotopes, particularly for use in nuclear reactors and weapons. Natural uranium contains a mixture of uranium 238 and uranium 235, with less than 1 percent of the latter. But U-235 is the isotope that undergoes fission, and so uranium must be enriched in this isotope (for certain reactors, to the 3 percent level; for explosives, to about 90 percent). Because the chemical reactions of the two isotopes are indistinguishable, they cannot be separated in the natural mineral by standard chemical means. A small difference exists, however, in the corresponding energy levels of atoms of the two isotopes; thus a slightly different light frequency is needed to excite the atoms from one level to another. Finely tuned laser light is discriminating enough to excite the atoms of one isotope and not those of the other. Further steps are needed to separate the excited atoms from the unexcited ones, and various techniques are being developed.

The significance of laser separation of isotopes is that it is cheap. The only large-scale method widely used heretofore is the gas diffusion system, in which vaporized uranium compounds are forced through porous materials. A single gas diffusion plant (the United States has only three) costs several billion dollars to build and a half billion dollars per year to operate; a laser system might cost one-thousandth of that. The high cost of isotope separation has been a major factor in preventing the spread of nuclear weapons to small nations. An inexpensive method would have significant geopolitical repercussions.

If the fission of uranium turns out to be unsatisfactory as a source of energy for the long term, one of the major alternatives being considered is *fusion*, the nuclear process in which hydrogen nuclei, as in the sun, fuse to form helium. This process occurs at extremely high temperatures, temperatures at which the system is a plasma—a gas of electrons and positively charged ions. The technical problem in taming the released energy is to control and confine this plasma. In the sun it is held together as a sphere by the huge gravitational field. In the laboratory scientists have sought to confine the plasma by surrounding it with intense magnetic fields that control the motions of the electrons and ions. A more recently introduced scheme has been to confine the plasma by striking the hydrogen fuel on all sides with intense laser beams.

The fuel is a minute pellet of deuterium and tritium[4] that absorbs laser energy, being thus heated to such high temperature that nuclear fusion commences. High-intensity pulsed lasers are used in this project, the objective being to produce more energy in fusion than is required to pump the lasers. Meeting this objective requires that the plasma be dense and that it remain confined for a sufficiently long time. In the laser system the density is extremely high, since the pressure of the beams compresses the initially solid pellet, and the confinement time needed is only of the order of 10^{-12} seconds. (The system is called *inertial* confinement, because the plasma is held together for this time only by its own inertia; that is, its own slowness in expanding.) In this time a mini-explosion would occur, the energy of which would somehow have to be captured as heat. A series of such explosions from successive pellets would provide a continuous and controlled heat source that could operate an electric power plant. At the present time it is not clear whether inertial confinement or magnetic confinement will turn out to be the preferred method for fusion, or even if either will lead to a practical energy source. In any case, implementation is probably three or four decades in the future.

The Hologram

Another laser-based device, probably rather far from its potential, is the *hologram*. We have seen that it is possible to record information about the *phase* of light from a particular source by creating an interference pattern between that light and light from a second source. Take as an example a single bright point sending out light in all directions. If that light is focused by a lens onto photographic film, there will be a single bright spot on the film—a record of the *intensity* of the source. In the absence of the lens, the light would strike the film all over the surface with various phases; but the film would not detect or retain this phase information. However, if the film is simultaneously illuminated with a plane wave (a broad beam of light with the same phase along each plane parallel to the film), then there will be recorded alternating rings of light and dark, corresponding to where the two beams were constructively and destructively interfering (fig. 9.7). This pattern is a complete record of the properties of the light that originally emerged from the point. In particular, if we later take the exposed film and pass through it a plane wave of laser light (coherent across the surface), the light that emerges will appear to diverge from a certain point behind the film—just the point where the original object was. In other words, there is an image of the point, but more than an image; we produce a set of rays of light that duplicates excactly the rays of light that emerged from the object point when the hologram was originally being made.

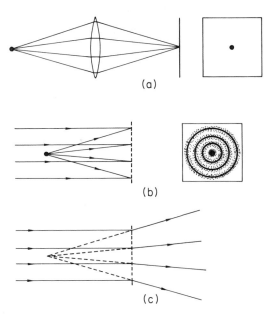

(a)

(b)

(c)

Fig. 9.7. The hologram of a point source. (*a*) Light rays from a point source are focused by a lens onto film and the image is a point of light. (*b*) Rays from the point source strike the film, and at the same time a series of parallel rays (which is equivalent to a plane wave) strikes the film. An interference pattern is created, a series of concentric alternating light and dark circles. This pattern, recorded on the film, is the hologram. (*c*) Illuminated by parallel rays, the hologram produces a set of diverging rays. The viewer on the right sees an image of the original point of light.

The recorded interference pattern is the hologram of a point source, a series of appropriately spaced light and dark rings. We can see that it behaves very much like a diverging lens in that it takes a plane wave of light and converts it to a diverging wave. Now suppose the source is not a point but a whole object, with light and dark regions reflecting laser illumination. The interference pattern between this reflected light and a plane wave from the same laser (the laser source is split into two parts, one going to the object and then to the film, the other going directly to the film) forms a hologram of the entire scene. The film will not look at all like the object (fig. 9.8) but will show a complicated array of interference fringes, swirling in many directions—sometimes like a fingerprint. When the film is lit by a laser source, the transmitted light is a duplicate of the light emitted originally by the object (fig. 9.9). And so when you view this light, you see all the properties of the object as if it were really there behind the hologram. Not only is it three-dimensional (possessing this property, as a real object does, because the light enters our two eyes from slightly different directions), but it also has parallax, meaning that you see

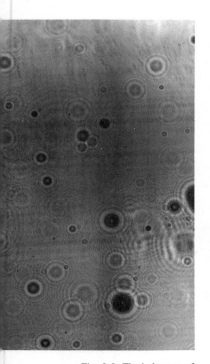

Fig. 9.8. The hologram of many fog droplets of varying sizes. Each droplet creates its own interference pattern, a set of concentric circles. Where these patterns overlap we see a complex arrangement of light and dark. In the hologram of a complete scene we would not recognize the scene, but the hologram would contain all information necessary to recreate the scene. (Courtesy Brian J. Thompson)

Fig. 9.9. In the upper figure a hologram is being made. A single laser illuminates the object and a mirror. On film we record the interference pattern of light from these two objects. In the lower figure the hologram is illuminated by parallel rays from a laser. Diverging rays proceeding from the hologram are a duplicate of the rays that would emerge from the illuminated object itself. (In addition to the virtual image shown here there is a second image, which is real. It is not seen by a viewer who looks at the virtual image.)

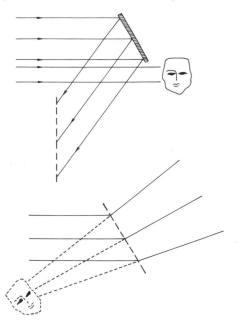

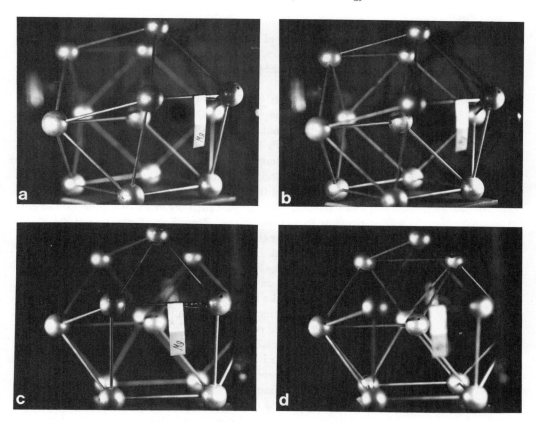

Fig. 9.10. A single hologram was made of this molecular model. Later the hologram was illuminated by a laser light, and that light was photographed. Note first that in photographs *c* and *d* the *viewpoint* is different from that in *a* and *b*. It is as if the viewer is further to the right in *c* and *d*, so that the label "Mg" is toward the front of the scene. The *focus* may also be varied. In *a* and *c* the front of the model is in focus; in *b* and *d* the front is out of focus, and a plane slightly farther back is in focus. (Courtesy Peter T. Gough)

a different image if you move your head from side to side (fig. 9.10). From the side you see one ear and one eye; from the front both ears and both eyes. The object of which the hologram is taken is effectively replaced by the hologram, and the latter can be stored on a shelf for subsequent examination. If the object has special optical properties, then its hologram also has these properties. A hologram of a magnifying glass will also magnify.

Since an interference pattern depends on the wavelength of light, the basic hologram is monochromatic. Color images can be made by creating several holograms at different depths in a thick photographic emulsion, each corresponding to a different wavelength. Later several monochromatic images may be combined, as in color film, to produce a multicolored image. A strip of holographic film in the form of a cylinder makes up a 360-degree hologram, one that can be viewed from front, back, and sides (not from top and bottom). Moving holograms are made of a series of holograms (several hundred), taken while the object is moving, in such a way that, when the observer moves his head from left to right in front of the image, the image moves—winking or blowing you a kiss.

While early predictions of widespread use of holography in television, movies, and other communication systems have not been borne out, numerous applications in advanced technology and research have been developed. Study of small three-dimensional systems under a microscope is hampered by the fact that only one plane at a time can be kept in focus. While the viewer is shifting from one plane of focus to another, the system (for example, a living one-celled organism or a collection of particulates in air) may move. Recording a hologram of the object allows all information about the object (for a given instant of time) to be stored. Later the hologram can be examined at leisure, and various planes brought into focus in turn. In high-speed (still) photography, when we do not know precisely where in the field the action will take place, a hologram taken with a short exposure time can store information about all points in the field. Later the camera is focused on the point where the action was. Optical instruments themselves, as mentioned above, can be replaced by a hologram, which is cheap (bypassing the expensive grinding of glass) and easily reproduced.

The hologram may become useful as a device for information storage based on the recording of light-carried signals, because as a record of incident light it differs in a fundamental way from an ordinary photograph. Each point on a photograph is a record of a particular point on the light-emitting object; but in holography each object point contributes to the entire interference pattern that is the hologram. Thus a small defect in a hologram does not destroy any particular part of the record (the record might be a page of digits—including your bank account) but merely makes the entire record very slightly less clear. Furthermore we can store a number of different records (pages) on a single hologram by making each record with the plane wave beam at a different angle. To read one of these records we only have to illuminate the hologram with light at the same angle.

Although limited in size, color, and mobility, holography has become a tool of the artist for the creation of ethereal, mysterious, and sometimes powerful images in space. Like computer music, it is a field where fruitful collaboration between scientists and artists has expanded the potential for esthetic experience. The Museum of Holography in New York City, for example, regularly shows new exhibits, abstract, whimsical, and horrible. A broken wineglass sits on a shelf; from a suitable distance we see an image of the unbroken glass—in the same place. A four-limbed creature sculpted by Aaron Kurzon has for its head a cylindrical hologram containing the image of a screaming man.

Information and Light Waves

There can be no doubt that the storage, processing, and transmission of information has become one of the major industries of advanced societies today, and more revolutionary developments based on waves of light are even now being planned in laboratories and boardrooms. The fundamental point here is that the *frequency* of light is several thousand times the highest frequency that can be produced electronically. And thus, as a carrier of information, light is thousands of times more effective than the electronic media used now in radio, television, and telephones. A wave "carries" information by the principle of *modulation*. For example, a radio wave with a frequency of 30 million hertz (cycles per second) can have superimposed on it a variation of its amplitude (fig. 9.11) with a frequency of 440 hertz (from the sound of the note A in the middle of the piano keyboard); it then transmits the signal to a point where a receiver and speaker reconvert it to sound.[5] This type of transmission is an *analog* process, in the sense that the modulated radio wave mirrors—or is analogous to—the original sound signal. More and more in modern applications signals are being processed *digitally:* the signal is converted into numbers, and the numbers are stored or transmitted. Since any number can be written with respect to base 2, data can be put in binary form, a series of 1's or 0's that is encoded by simply turning the radio signal on or off. Digital transmission is still a form of amplitude modulation, where the modulating signal is either 1 or 0 during a series of fixed time intervals. But for either analog or digital transmission the time variation of the modulating signal must be very much slower than that of the carrying signal—because if it is not, then the shape of the modulating signal, which is what is to be read at the end of the transmission line, will be affected by the shape of the carrying signal. In the digital case, the time interval during which one sends a single 1 or 0 must contain many oscillations of the carrier. It follows that if the carrier is of a higher frequency, the digital time interval can be shorter, and more digits can be transmitted per second. Light with frequency around 5×10^{14} hertz can carry information at a rate more than 10 million times greater than typical radio waves and thousands of times greater than microwaves.

This is the objective and it has called forth, and is still calling forth, a series of ingenious inventions designed to implement a light-carrying system. For a laser beam cannot simply be sent through the air. It is stopped by fog or a cloud. The key to practical light wave transmission has been the successful development

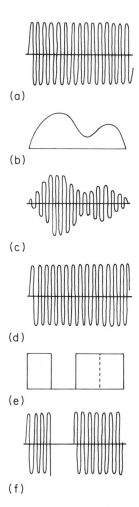

(a)

(b)

(c)

(d)

(e)

(f)

Fig. 9.11. The propagation of signals. (*a*) A high-frequency carrier signal, electronic or optical; (*b*) a low-frequency analog signal, sound waves, for example; (*c*) the carrier signal is modulated by the amplitude of the analog signal; (*d*) the carrier signal; (*e*) a digital signal with the information "1-0-1-1"; (*f*) the carrier signal modulated by the amplitude of the digital signal.

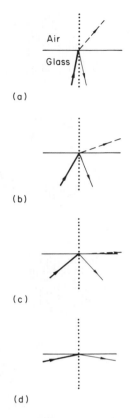

Air

Glass

(a)

(b)

(c)

(d)

Fig. 9.12. Reflection and refraction of an incident light ray (*heavy line*) traveling in a dense medium (glass) and entering a less dense medium (air). The refracted ray (*dashed line*) is bent away from the normal to the interface. The reflected ray (*solid line*) makes the same angle with the normal as does the incident ray. As the incident ray moves closer to the interface (cf. *a* with *b*), away from the normal, the refracted ray moves still closer to the interface. A critical angle is reached (*c*) where the refracted ray is parallel to the interface. For a more grazing incident angle (*d*) there can be no refracted ray. Therefore only reflection occurs; light does not leave the glass.

of a low-absorption *light fiber,* a thin flexible glass cylinder, a fraction of a millimeter in diameter, through which light can travel a distance of at least several miles without significant loss of intensity.

Light travels in the fiber, even bending with it around corners, by the action of an elementary and long-known principle, *total internal reflection* (fig. 9.12). We recall the laws of refraction and reflection: when a light beam reaches an interface between two different media, there is both a reflected wave and a refracted wave. If the beam goes from a region of high index of refraction (such as glass) into a region of lower index of refraction (air), then the refracted beam is bent away from the normal to the interface; the reflected beam of course returns into the glass. As the incident beam becomes closer and closer to the interface, a point is reached where the refracted beam emerges just parallel to the interface. This is the limit of refraction into air. For an incident beam still closer to the interface there cannot be any refraction, and so the light is entirely reflected. Moreover, it is reflected into the glass making the same angle with the interface as did the incident beam. If the glass has two parallel edges (as in a cylinder or a slab) the beam continues in a zigzag path, never escaping into the air. And, provided the glass cylinder is not bent too sharply, light will also be totally reflected—back into the glass—in the bent regions; so light travels around corners (fig. 9.13).

Light fibers have been used in various applications for quite a few years: for example, in the endoscope, a device that permits doctors to examine and illuminate interior portions of the body by passing a light fiber through the esophagus or the rectum. Dentists use a light fiber to bring light along the arm of the drill and cast it directly on the tooth being worked on. These are short fibers in which the loss of intensity is not significant. The development in 1974 of fibers capable of carrying signals over several miles opened up the possibility of light wave systems not only for telephones but for numerous new or planned systems requiring high-density information transmission: for example, interactive television, picture phones, and computer-supported networks supplying data for financial and technical operations.

Light wave telephone systems have been tested and installed in dozens of major cities in North America, Europe, Japan, and China, and many systems involving broader data services are already in operation. High-density data handling is most necessary in cities, where the density of consumer and business transactions is greatest and space is at the greatest premium. Intercity communications via light wave was inaugurated in 1983, using booster stations at intervals of about 20 miles to reconstitute a weakening signal. Since light signals and electronic signals can be interconverted, it is not necessary to replace the entire telephone system

Fig. 9.13. Light traveling in a fiber. Light first strikes the glass-air interface at a fairly shallow angle and so is entirely reflected. No light is lost via refraction. If the fiber is curved (but not too sharply), light will strike the interface a second time also at an angle beyond the critical angle. So again it is totally reflected. Light continues in this zigzag manner, without being refracted into the air.

with a light wave system, but light wave pieces can be added one by one to the existing system. There is even a plan for a light wave transatlantic cable. Present-day research centers on the need for new types of glass for fibers that can carry signals hundreds of miles without significant attenuation.

Along with fibers as transmission lines and lasers as light sources, a complete new technology is being created to build miniaturized integrated optical circuits in a manner analogous to the building of integrated solid state electronic circuits. Light moves in a thin crystalline film, in which may be embedded prisms and lenses, switches to modulate the signal, minute semiconductor lasers, or light-emitting diodes. (These are light sources, cheaper than lasers, which operate essentially as reverse photovoltaic cells, converting electric current into light. They are sometimes used for the readout in calculators.) Light in the film moves in the same zigzag manner as in the fiber, and hence its effective speed depends on the angle of the zigzag. By means of prisms light can be inserted into the film at different angles, and beams with different entrance angles act independently in the film. Each beam can carry a separate signal, and thus a single film carries information on many different channels simultaneously.

The index of refraction of a crystal is changed by an applied electric field, and this effect can act as a high-speed switch. It bends light in one direction when the field is on, in another direction when the field is off. Corrugations at the interface of a film act as a diffraction grating, bending light in different directions according to its wavelength. This provides a method of separating signals that come in simultaneously on the fiber on several different carrier frequencies; it acts like the tuner in a radio receiver. Since the entire optical system is built in a crystal, it can be integrated with electronic components by doping the crystal where needed.

An important feature of light wave communication systems is that, unlike electronic systems, they are not disrupted by electromagnetic disturbances, such as those due to large-scale electrical machinery, geomagnetic storms, or power lines. There is also an intense electromagnetic pulse associated with the explosion of nuclear weapons, an effect that could obliterate much of the electronic communication and data processing on which military responses depend. For this reason, an effort is under way to substitute light wave systems for electronic systems.

Integrated optics is a new and rapidly expanding field, and the systems that will finally be in place ten years from now may be quite different from what we imagine at present. But we may perhaps briefly look a bit further into the future at other technologies that may be arriving before too long. If light is a powerful tool because of its high frequency, can we not expect that electromagnetic radiation of still higher frequency will be powerful in new ways? One of the by-products of research in high-energy physics has been the availability of controlled *synchrotron radiation*—a certain kind of electromagnetic wave in the far-ultraviolet and x-ray regions, from about 100 nm down to 0.01 nm. This radiation is produced by accelerated charged particles (usually electrons) moving in circular orbits. Originally electron accelerators were built for basic research in nuclear and particle physics; but more recently some have been adapted for the study and use of synchroton radiation, and new machines have been built for this purpose alone.

As a research tool synchrotron radiation opens up several new scientific fields, particularly in the study of crystal structure and of atomic arrangements near surfaces. Experimental characterization of surfaces is important in analyzing phenomena such as chemical catalysis, corrosion, and the electrical properties of *p-n* junctions.

While it is not now being proposed to use synchrotron radiation to carry information, we may soon see its use as a substitute for light in the production of microelectronics on silicon chips. Miniature circuit elements are produced (to summarize briefly a highly intricate technology) by a photolithographic method. A thin semiconductor film is covered by a mask outlining the region where a *p*-type layer is supposed to remain. It is then exposed to ultraviolet light, which induces a molecular change in the uncovered portions of the layer. The film is then etched with a chemical that dissolves the unexposed portions of the layer, leaving a pattern that duplicates the pattern of the mask. For very small structures the method is limited by the wavelength of light, since light creates a diffraction pattern below the mask rather than a sharp shadow. But as we go to shorter wavelengths, diffractive effects become less important. Although many technical problems remain to be solved, lithography with x-rays may make it possible to reduce the size of electronic components from the present minimum, around 2,000 nm (or 0.002 millimeters), to 100 nm. We reach here a dimension that corresponds to a square with only 1,000 atoms along the sides, a million atoms altogether. Impressive as the present generation of microelectronic devices is, we are really only at the beginning. The ultimate limit in miniaturization is the single atom itself.

Relativity

10

When we look back on a major scientific advance, we often see it as both an ending and a beginning. It subsumes the tentative striving of earlier generations, resolves paradoxes, creates elegance, and embodies truth. Yet these very accomplishments clear the way for new questions posed at a deeper level and new paradoxes, which can be resolved only by further advances in the next generation. So it was with Maxwell's theory of electromagnetism. The interaction of electricity and magnetism as studied by Faraday and his predecessors, the wave properties of light as seen in interference and diffraction, the finite velocity of light in vacuo and its bending in air and glass, the extension of the spectrum to ultraviolet and infrared, all these phenomena are accounted for by electromagnetic theory. But even as it was enunciated in 1862, fifteen years before Einstein's birth, its mathematical equations contained the seeds of the theory of relativity.

The Luminiferous Ether

The luminiferous ether lay at the foundation of the structure Maxwell had created. His accomplishment, in the nineteenth-century view, had been to unify the ethers of light, radiant heat, electricity, and magnetism. And though the ether's properties of tenuousness and rigidity seemed paradoxical in comparison with the properties of ordinary matter, the ether nonetheless was able to provide a consistent basis for electromagnetism. In Maxwell's words, "Whatever the difficulties we may have in forming a consistent idea of the constitution of the ether, there can be no doubt that the interplanetary and interstellar spaces are not empty, but are occupied by a material substance or body, which is certainly the largest, and probably the most uniform body of which we have knowledge." [1] But before long a new paradox arose, connected not with the constitution of the ether but with its location in space; specifically, it concerned the elementary question of what happens when the earth in its circular orbit moves through space and hence through the ether.

It seemed reasonable to suppose that there would be created an "ether wind" as the earth moved, and numerous experiments, covering the half-century from 1880 to 1930, were carried out to detect this wind. The ether can be thought of as a frame of reference relative to which the speed of light is given. For example, when we say that sound travels at 344 meters per second, we mean that the wave moves at that speed relative to the air, and when we say that light travels at 300,000 kilometers per second, we mean this speed relative to the ether. Air is the medium of sound; ether is the medium of light. Light emitted by a source, say, a star fixed in the ether, then travels toward us with the speed 300,000 kilometers/second. But if the earth is moving toward the star, then *relative to earth,* where the light is detected, the speed of light must be greater. Similarly, if the earth moves away from the star, the speed of light reaching us must be smaller. Thus we envisage an experiment based on the passage of light in different directions, to detect the motion of the earth through the ether.

The most famous attempt at such a measurement was the *Michelson-Morley experiment,* designed by the American physicist Albert Michelson (fig. 10.1) and first carried out in 1881. Michelson used a terrestrial light source in a device he had invented called an *interferometer* (fig. 10.2). As its name implies, the interferometer measures interference—interference between two light beams that derive originally from the same source (hence are coherent). The initial beam strikes a glass plate at a 45-degree angle; part of the beam is reflected, part transmitted. The two sec-

Fig. 10.1. Albert Michelson (Michelson Museum; courtesy AIP, Niels Bohr Library)

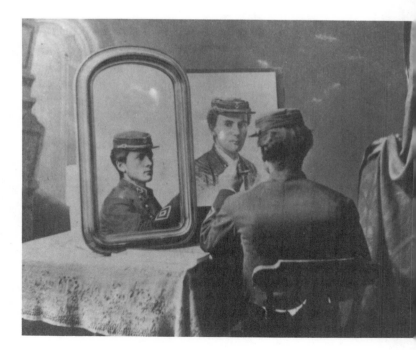

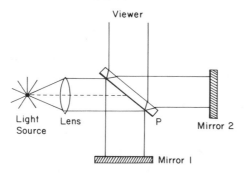

Viewer

Light
Source   Lens   P   Mirror 2

Mirror I

Fig. 10.2. The Michelson interferometer (slightly simplified). Parallel rays of light produced by a lens strike plate *P* at a 45-degree angle. This plate is what is called a half-silvered mirror: it splits the beam into two parts, one reflected downward, the other transmitted to the right. The first part is reflected upward by mirror 1, passes through *P*, and continues upward toward the viewer. The second part travels to the right, is reflected by mirror 2, and is then reflected from *P* upward toward the viewer. The recombined beams are viewed through a telescope in which interference effects may be observed.

ondary beams travel at right angles and are eventually brought together by means of further mirrors and are viewed through a telescope. Since the two beams have traveled different paths—perhaps one has traveled parallel to the earth's motion and the other perpendicular to the earth's motion—the viewer sees an interference pattern, alternating fringes of light and dark. This is not surprising; but the final step in the experiment is to rotate the entire apparatus through 90 degrees, effectively interchanging the paths of the two beams. Now we expect to see a shift in the fringes since the light paths through the ether have changed. The result of the Michelson-Morley experiment was negative; no fringe shift was observed. Although Michelson regarded his original experimental attempt as a failure, more and more precise results obtained over several decades confirmed this crucial null result. It was as if the ether, a cosmic medium extending out to the stars, remained stationary with respect to this minor planet earth in its annual circuit about the solar system.

More reasonable was the hypothesis put forward by Fresnel, that while the ether may be stationary in interstellar space, in the vicinity of earth it is dragged along with the moving planet, creating a kind of ether atmosphere. Just as our gaseous atmosphere moves with us, so does the ether, in such a way that experiments (such as the Michelson-Morley experiment) performed within the confines of the atmosphere do not detect the motion of the earth through space. The *ether-drag* hypothesis, however, foundered upon another observation, this one involving light reaching us from interstellar space. In 1728 James Bradley (fig. 10.3) had observed that the position of a star moves through a small circle (or ellipse) during the course of one year's observation through the telescope, a motion known as the *aberration of starlight*. Bradley reasoned that this was an apparent motion, resulting from the changing velocity of the earth in its orbit. The effect is analogous to trying to catch vertically falling raindrops in a tube (fig. 10.4). If the tube is stationary, drops fall down the center of the tube and reach a receiver at the bottom. But if the tube is being transported

horizontally, a raindrop that falls into the mouth of the tube will not reach the bottom; it strikes the side of the tube on the way down. To allow drops to reach the bottom, the tube must be tilted forward (the angle depends on the speed of the tube and that of the falling drops). Similarly, the tube of a telescope must be tilted forward to catch the falling beams of light when the tube is moving horizontally—carried by the moving earth. When the astronomer finds that he has to tilt the telescope in different directions, he interprets this as apparent motion of the star. Thus was aberration of starlight observed and explained by Bradley.

But stellar aberration is inconsistent with the ether-drag hypothesis because the motion of light is defined relative to the ether. If the ether moves with the earth, then when light enters the ether the beam is dragged along with the earth, as is the telescope, and so finally the direction from which the light enters the telescope is not affected by the motion of the earth through space; there would be no aberration.

Thus the ether-drag hypothesis is untenable, as is the hypothesis of a cosmic stationary ether. A fundamental problem emerges, as was recognized by the physicists of the late nineteenth century, which we may express by the question, Where is the ether? or, more straightforwardly, How fast does light really go? For although the speed of light has been accurately measured in many experiments (see app. I), the question here is how is that speed affected when the source of light or the receiver of light is moving. Much theoretical speculation on this matter occurred in the years after 1887, when a second and more precise version of the Michelson-Morley experiment caught the attention of the

Fig. 10.4. Aberration of starlight. A vertically falling light beam enters a telescope tube that is traveling (because it is mounted on the moving earth) to the right. If the telescope is vertical the beam does not reach the bottom where the viewer is; the star is not seen. By tilting the telescope to the right we allow the beam to reach the bottom. We interpret this result as if the star were located at an angle toward the right.

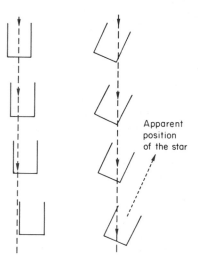

Apparent
position
of the star

Fig. 10.5. *Left to right:* Albert Einstein, H. A. Lorentz, and Arthur Eddington. (Courtesy AIP, Niels Bohr Library)

scientific establishment. And in 1905 Einstein (fig. 10.5) found the solution lying at the heart of Maxwell's equations for electromagnetism.[2]

Einstein's Relativity

Maxwell's equations are certainly consistent with the idea of *relativity,* meaning simply that things, certain physical quantities, are relative; they have different values when viewed relative to a fixed or a moving frame of reference. For example, a fixed electric charge will set up an electric field, but a moving electric charge will set up both an electric and a magnetic field (as in Oersted's experiment in which an electric current, really nothing but moving electric charges, acts as a magnet). If we observe the moving electric charge from a frame of reference moving with it, then the charge appears stationary; there is no magnetic field. Is there a magnetic field or isn't there? The answer is simply relative. There is in one frame of reference, and there isn't in the other frame of reference. This conclusion is not really so puzzling: motion itself is relative. Is the charge moving or is it not? Again, the answer is relative. Some quantities vary with the frame of reference and some do not. Since the *laws* of physics cannot depend on who is observing them, these laws must deal with quantities that do not vary with frame of reference—or the laws must relate quantities that vary with frame of reference in the same way. This aspect of relativity, although emphasized by Einstein, was not new to science and had been understood by the classical physicists since Galileo and Newton.

Any quantity that is *not* relative (that does not vary with frame of reference) is called an *invariant.* Einstein's new step is to assert that the speed of electromagnetic waves, the speed of light, is one of the invariants. Light waves are propagated because of the dynamic interactions between electric and magnetic fields, and Maxwell's theory draws upon experiments with these fields to predict that light travels[3] at a definite speed, c (to use the universal symbol for this fundamental quantity)—not c in some frames of reference and some other number in other frames of reference, but simply and solely c.

Viewed in traditional terms this is an almost incredible conclusion. For it violates intuition and experience, that when something moves toward us, a baseball for example, it reaches us more quickly if we move forward to meet it; that is, its speed is greater relative to the moving receiver. (Suppose the ball approaches at 90 miles per hour, and I run toward it at 15 miles per hour. It hits my glove at 105 miles per hour. This simple *additive* law is known as the principle of Galilean relativity.) We are left with only two alternatives. One is to say that the predicted velocity of light is meant to be relative to a particular and very special frame of reference. The second alternative is to accept the speed at c in every frame of reference, accept its violation of our intuition, but recognize that our experience really does not extend to objects moving at such high speeds. The first alternative was that taken by the nineteenth-century theorists; and the ether, besides being the "substance" that carried the wave, had the function of defining the special frame of reference. The Michelson-Morley result shows that it is impossible to determine the presumed fixed position of the ether. Thus, Einstein's relativity chooses the second alternative, rendering the ether an unnecessary construct in the theory.[4]

How can the additive law of relative velocities be rejected without violating commonplace experience? Again, the new theory does not "overthrow" the old. Rather, the old theory must continue to hold approximately in a limited domain, and the new theory extends our understanding to a wider domain. In this case there must be found a new way to transform velocities from one frame of reference to another, with the properties that (1) when all speeds are much less than the speed of light, the additive law of Galilean relativity is in error by a slight (often unmeasurable) amount, and (2) the law changes in a continuous way as we approach higher speeds, so that for an object traveling at c itself, the speed is the same in all frames. Among other things, we are forced to the conclusion that the measured values of time, as viewed in two different frames of reference, may differ. (For example, two events may be simultaneous in one frame of reference and not simultaneous in another.) The crisis of the ether forces us to choose between Maxwell's theory of light and conventional concepts of space and time. We (or rather nature) finally choose light.

Mass and Energy

Relativity enters physics at such a deep level that its implications are felt in every part of the subject. To explore all these ramifications would take us too far afield, but one implication is crucial to subsequent developments in the theory of light. This is the connection between mass and energy. In classical physics an object possesses mass because it is made of matter; an object possesses energy because it is in a state of motion. Mass is the resistance an object exhibits when we try to push and accelerate it; mass inheres only in the atoms or molecules that make up the object. When we do succeed in accelerating the object, we have supplied it with energy. Thus energy would be a relative quantity, different when seen in a moving frame of reference; mass would not. Maxwell's electromagnetism compels a new view of these questions. For suppose we try to push an electron. As the particle gains speed it also begins to set up in the surrounding space a magnetic field. This field is a form of energy, in addition to that of the moving electron. In consequence, we have to push harder to accelerate the electron; to supply the additional energy going into the field, we must push harder than would be necessary if the electron were not electrically charged. It is as if the electron had an additional mass, resistance to acceleration, because it participates in the phenomenon of electromagnetism. Moreover this additional "electromagnetic" mass is greater the higher the electron's velocity. Physicists at the turn of the century speculated that in the case of the electron its *entire* mass might be ascribed to this electromagnetic effect; and observations of high-speed electrons indicated that the electron's mass did indeed increase in just the way predicted by electromagnetic theory.

As an electron slows down, both its energy and its mass decrease, until finally when it is at rest it possesses a certain characteristic minimum mass (9.11×10^{-28} grams). We may argue then that even at this stage the electron has energy, a new form of energy associated with its internal structure alone, which we may call its *rest energy*.[5] One reading of the equation $E = mc^2$ is as the relation between E, the rest energy of a particle, and m, the mass the particle has when at rest (called the *rest mass*). When the particle is moving, both E and m increase, but they remain connected by this fundamental equation and the universal constant c^2. The theory of relativity, in 1905, showed that the increase in mass with velocity is to be expected on general grounds as a consequence of the new transformation principles of space and time; it applies not only to the electron but to any particle, any matter, whether electrically charged or not.

A particle's rest mass thus embodies a new form of energy. It takes its place under the umbrella of the principle of conservation of energy, which states that in an isolated system the totality of all forms of energy (such as heat, radiation, and energy of motion) must be unchanged, although one form of energy may be transformed into another. Under certain conditions a particle's rest energy may disappear, with the consequence that energy in some other form takes its place. The particle has ceased to exist. Conversely a particle may be created by assembling its rest energy at the expense of energy in another form. The deeper question of *why* we have in our universe the particular particles we do (the electron, the proton, etc.), with the rest masses which, to some extent, define them, remains unanswered.

The increase of a particle's mass with velocity has been verified repeatedly since 1901, but it should be stressed that at ordinary velocities the effect is extremely small. The mass is given by multiplying the rest mass by a factor (called *gamma*) that is very slightly greater than 1; for motion at speeds small compared with the speed of light the change is not detectable. On the other hand, if an object travels at a speed close to that of light, gamma can be considerably greater than 1, and as the speed approaches closer and closer to the speed of light, gamma becomes larger and larger, without limit. (Mathematically we say that gamma is infinite when the speed equals c.) Here is the basis for the often-quoted statement that nothing can go faster than the speed of light. For mass is resistance to acceleration, and so as we push an object close to the speed c, its resistance to that push becomes stiffer and stiffer. It becomes the "immovable object." The speed of light is never achieved, but only approached.[6]

Where does that leave the photon? The particle of light certainly has energy. Quite literally we "see" its action upon molecules and atoms; we understand how the photon's energy depends on the light's wavelength. Given energy, the photon also has a mass, as determined by $E = mc^2$. How does the photon reach the "speed of light" without its mass becoming infinite? The answer to this question points to the special part played by the photon in the panoply of modern physics. For (almost) uniquely among nature's elementary particles, the photon has a rest mass of precisely zero. The mass of the photon is given by multiplying its rest mass by the factor gamma, mentioned above. But for the photon gamma is infinite. Thus we have infinity times zero, a product which mathematically is said to be "undefined." In a physical sense the product may be well-defined and finite, and thus it is for the photon.

The photon has mass and energy, although no rest mass. One of the properties of its mass, predicted later as part of Einstein's theory of general relativity, is its response to the force of gravity. Like the earth, the photon should travel a curved path in the presence of a strong gravitational field like that of the sun. This prediction was verified spectacularly in observations made in 1919, during a total eclipse of the sun. With the sun's light blocked, astronomers could determine the precise location of stars near the sun's rim. A change in the *apparent* location of these stars, due to bending of light rays, confirmed the gravitational response of the photon.

As we move from the realm of ordinary velocities to that of velocities close to c, we find ourselves in a situation for which our experience has not prepared us. The meanings of physical terms change and we can rely only on operational definitions. Gravity acts on the photon but cannot make it go faster or slower. As nature's most subtle creation, the photon teeters on a knife-edge. With rest mass equal to zero, if it traveled the slightest bit slower than c, its energy and mass would become zero. It would have ceased to exist. Alternatively, if the photon traveled at c but had a rest mass the slightest bit greater than zero, it would have infinite energy and, in effect, could never have been created.[7]

Quantum Electrodynamics

The quantization of light, along with the theory of relativity, forces upon us a new perspective concerning this number we call c. It is no longer just the speed of light but is an inborn characteristic of the universe, of spacetime if you will; it is the speed greater than which nothing can go. Within spacetime are created, for reasons we do not know, certain fundamental types of particles. Each is defined by a small set of parameters, one of which is rest mass. (Another is electric charge; a third is intrinsic spin.) The photon is one, perhaps the only one, with zero rest mass. Hence it may (and it must) travel at c.

But in another sense relativity asks us to see that all the elementary particles behave in a way that we would earlier have thought to be characteristic of only the photon. They can be created and destroyed. The electron, the proton, the neutron, the muon, the pion (the muon and pion are particles discovered in 1936 and 1947, respectively, both having a mass greater than the electron's but less than the proton's), and the neutrino—none are permanent. Just as the black ink of this page gobbles up photons—destroys them—so will a neutron, broken free from a nucleus, spontaneously disappear in a matter of minutes and so will a proton and an antiproton annihilate each other. Just as the

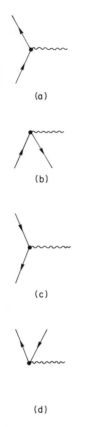

(a)

(b)

(c)

(d)

Fig. 10.6. Fundamental vertices of quantum electrodynamics. A line with an arrow pointing upward represents an electron, a line with an arrow pointing downward a positron, and the wavy line a photon. (a) An electron emits or absorbs a photon; (b) an electron and a positron annihilate each other and produce a photon—or an electron, a positron, and a photon all unite and are destroyed, leaving a vacuum; (c) a positron emits or absorbs a photon; (d) a photon disappears leaving in its place an electron-positron pair—or out of the vacuum are created an electron, a positron, and a photon.

quivering atoms in the filament of an electric light bulb send forth new photons every second, so will a shower of pions be created in a proton-proton collision engineered in a high-energy accelerator.

So readily are elementary particles created and annihilated that we are tempted to conclude that all is impermanence in this none too solid world. Yet nature is far from wanton; for there are rules that set strict boundaries around the transformations of the subatomic domain. These rules are the conservation laws: in the creation and destruction of individual particles certain total quantities cannot change. One such quantity is energy. When a particle is destroyed, the energy that was associated with its rest mass must appear somewhere else, sometimes as the rest mass of a newly created particle, sometimes as the energy of motion. Another conserved quantity is electric charge. We cannot lose a proton unless its positive charge reappears attached to some new particle (or unless a negatively charged particle is simultaneously extinguished, thereby balancing the lost positive charge.) Elementary particle transformations are governed also by other, somewhat more abstract conservation rules that we will not elaborate further. But the principle of creation and destruction of elementary particles, emerging from relativity, leads directly to the theory, formulated in the late 1920s, that became the most advanced and complete theory of light, and subsequently a model for all future theories of the subatomic world.

It was called *quantum electrodynamics,* because it grew from the amalgamation of the quantum theory of Schrödinger and Heisenberg with Maxwell's field theory of electrodynamics. It also incorporated a third element, the theory of *antiparticles,* put forward in 1928 by P. A. M. Dirac (see fig. D.1) and confirmed in 1932 by Carl Anderson's discovery of the positron, antiparticle to the electron. The antiparticle is an object that has the same rest mass as its corresponding particle but is in all other respects opposite. The positron, for example, has a positive electric charge; when brought into proximity an electron and positron annihilate each other, leaving, as required by relativity, the appropriate amount of energy in another form. Quantum electrodynamics was first a theory of electrons, positrons, and photons, although it applies very generally to the interaction of light with any electrically charged particles. It will be easiest, however, for us to discuss the theory in connection with electrons and positrons, because for these particles the electromagnetic interaction is the dominant one. (For protons and antiprotons there is not only an electromagnetic interaction but also a nuclear interaction—discussed below—which is stronger than the electromagnetic one and in some situations swamps it.)

The core of the theory is a three-pronged *vertex,* a point in space and time at which two electrons and one photon meet. We can think of this as a process in which one electron arrives and is annihilated, and at the same instant a photon and a new electron are born (fig. 10.6a). Equally well, the photon in the vertex could be the particle destroyed. Moreover, since a positron is an electron's opposite, the creation of an electron is closely related to the destruction of a positron. Therefore the primitive vertex is understood to include a process (fig. 10.6b) in which an electron and positron are both *annihilated* and a photon is produced (or all three are annihilated, leaving nothing but a vacuum).

Perhaps it is apparent at this point that the primitive vertex is not a process that occurs by itself in nature.[8] Rather it is a building block out of which the real events of the world are generated, by combinations of two or more primitive events. As an example, consider figure 10.7a. A photon on the left is destroyed, creating an electron and a positron. The positron travels for a short time, meets an electron on the right, and the two annihilate to produce a new photon. The net effect is that an electron and a photon have come together and an electron and a photon emerge. It is the process of "scattering" of light by matter. In this case the matter is a single electron, but a similar construction accounts for scattering of light by atoms and molecules, and ultimately accounts for reflection, refraction, and the blue of the sky.

Figure 10.7b shows a process in which two electrons interact and emerge with a newly created photon (just the event that cannot proceed with only one electron). Figure 10.7c shows a process in which there is no apparent photon at all. One electron enters the scene and emits a photon, which travels across to where it is absorbed by a second electron. Imagine two rowboats coasting side by side on a lake. A person throws a medicine ball from

Fig. 10.7. Diagrams that are built out of the fundamental vertices of quantum electrodynamics and that represent real processes. (*a*) A photon and an electron scatter; (*b*) two electrons interact and emerge along with a newly created photon; (*c*) two electrons scatter, the interaction mediated by an exchanged virtual photon. Diagrams of this type were first drawn by Richard Feynman as a way of summarizing the detailed procedures for calculating the probabilities for these processes.

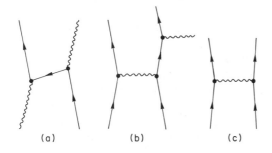

(a)   (b)   (c)

one boat and it is caught by someone in the second boat. The second boat swerves to the side under the impact of the caught ball. The first boat swerved in the opposite direction when the ball was thrown. Viewed from above, and if we could not see the exchanged medicine ball, it would look as if some mysterious force had driven the boats apart, as if a spring between them had been released. Two electrons come together, transfer a photon from one to the other, and then swerve apart, *as if there had been a force between them*. The exchange of a photon *is* the repulsive force between two electrons.

Here lies the power and eloquence of the modern theory of light, molded by quantum mechanics and relativity. For quantum electrodynamics offers in the truest sense of the word an *explanation* of the forces of electricity and magnetism. They are derived effects, constructed out of the primitive three-pronged vertex that unites the photon and the electron. A similar vertex exists between the photon and any electrically charged particle (and also some electrically neutral particles, like the neutron). And the combinations and permutations of these vertices account for electromagnetic interactions of all kinds. Thus quantum electrodynamics lies behind the very existence of atoms (binding them together), the formation of solids, liquids, biological molecules, and life.

The primitive vertex implies that the potential for production of photons, and electrons and positrons as well, exists at all times as an intrinsic factor in our universe; all that is needed is some source of energy. Thus the early universe, compressed and hot after the big bang, was flooded with these particles. From a time just a few minutes (according to theory) after the big bang until the universe reached the age of several million years, it was dominated by radiation, in the sense that the quantity of energy in the form of radiation far exceeded that in the form of matter. This period, known as the *radiation era,* must have been preceded by an era in which the dominant form of energy was electrons and positrons in approximately equal number. Such electron-positron pairs would have been copiously created at the even more intense energy levels of the first few minutes. Following the radiation era is the matter era, in which we find ourselves now. The dominant form of energy is the rest energy of matter (perhaps hydrogen atoms like those in the stars, unless the "invisible mass" takes some more exotic form), and the residual radiation is what we see as the cosmic background radiation left over from the big bang.

A word of caution is in order before we leave the subject of the exchanged photon, like the one in figure 10.7c. For unlike the medicine ball in our analogy, the exchanged photon is in a certain sense not real. We have seen that a single electron cannot spontaneously emit a photon; and so the process envisaged in the diagram is real only in its entirety, only with the photon absorbed as well as emitted. The existence of the photon is transitory, like the flash of Tinker Bell, a violation of conservation of energy for such a short period of time that we are, in principle, unable to catch it.[9] Such photons, which appear only at intermediate points in a process, are called *virtual*. Although we do not see them, we know that they play this intrinsic role in electromagnetic phenomena because photons do appear in processes like that of fig. 10.7b. Two electrons interact (exerting forces on one another) *and* emit a photon in the process. It is indeed the quantum representation of the vibrating (or otherwise accelerated) electric charge that emits light according to Maxwell's classical theory. The electron that emits light in figure 10.7b does so only because it interacts with the second electron; in classical language this means it is acted upon by a force and so is accelerated.

Quantum electrodynamics was not developed overnight. It began in the 1920s in the hands of Heisenberg, Dirac, and some of the other quantum revolutionaries, but was not fully realized until the 1940s and the work of Julian Schwinger, Richard Feynman, Sin-itiro Tomonaga, and others. By the 1950s it was recognized as the most advanced and complete theory of matter and radiation, capable of accounting for the widest variety of phenomena and making predictions which have been verified to a high degree of accuracy. More than just a theory of electromagnetic interactions, however, quantum electrodynamics introduced a new way of thinking which, through the present day, has come to dominate theories about all forms of interaction.

Fundamental Interactions

Twentieth-century physicists have distinguished, in addition to electromagnetism, three other fundamental interactions that seem to be necessary to describe nature. One is gravitation, the attractive force between all objects possessing mass. A second is the *strong force* (also called the nuclear force), which acts among protons and neutrons, providing the glue that holds them together in nuclei. And the third is called the *weak force,* associated among other things with beta decay, the type of radioactivity in which a nucleus emits electrons. The success of quantum electrodynamics suggested that mechanisms of a similar type, involving the exchange of virtual particles, are at work in all other interactions. The idea is basically a generalization of the Faraday-Maxwell concept of a *field;* that is, that interactions do not occur because two particles "touch" or because they act on each other instantaneously across a distance, but rather they occur because each creates around itself a field, a region of influence, that in time reaches and influences the other particle. The twentieth cen-

Fig. 10.8. Hideki Yukawa
(Courtesy AIP, Niels Bohr
Library, W. F. Meggers
Collection)

tury adds to this concept the quantization of that field, that just as light in a sunbeam is packaged in discrete units (real photons), so the field that mediates electromagnetic forces is quantized (virtual photons); and similarly the fields that mediate other interactions are quantized, incorporated not in photons but in particles of a similar nature.[10] Modern theories of elementary particles have thus come to be called *quantum field theories.*

The first proposal along these lines was made by Hideki Yukawa (fig. 10.8) in the 1930s in connection with the strong force. Yukawa recognized that the most significant difference between the strong force and the electric force was a difference in range. The strong force extends over a distance of only a few times 10^{-15} meters (about the size of a nucleus), and beyond that range it becomes essentially zero. Electric (and magnetic) forces also become weaker when two interacting particles move apart, but there is no characteristic distance beyond which they vanish. Yukawa saw that the range of the force was associated inversely with the *rest mass* of the exchanged particle: a short-range force would be associated with the exchange of particles having non-zero rest mass, whereas a long-range force (like electromagnetism) would be associated with a massless particle. (See app. J.) Thus he predicted the existence of a particle whose rest mass would be 200 or 300 times that of the electron and which would interact strongly with protons and neutrons. Such a particle was the *pion,* discovered a decade later. Possessing some of the properties of the photon, the pion comes in three varieties, with positive, negative, and zero electric charges. And while we do not see the virtual pions, which mediate the nuclear interaction, we do see events in which two protons collide and scatter, creating in the process one or several pions, in close analogy with the production of light during the electromagnetic interaction of two charged particles.

Similarly a quantum field theory of gravitation is based on the exchange of a particle, the graviton. Since the gravitational force is long range like electromagnetism, the graviton has zero rest mass. It is believed that production of free gravitons occurs, but because the interaction is extremely weak, this analog of the photon has never been detected.

The principle of describing the interactions of elementary particles in terms of exchange of particles lies behind all contemporary approaches to unified field theories. Physicists seek to create more comprehensive theories that will encompass and unify two or more of the four interactions. This objective was long associated with the name of Einstein, who sought unsuccessfully a theory to unify electromagnetism and gravity (as Maxwell had unified electricity and magnetism). The most important recent development in this direction has been the emergence of a theory,

Fig. 10.9. *Left,* Abdus Salam (Courtesy AIP, Niels Bohr Library, Marshak Collection); *Right,* Steven Weinberg (Courtesy AIP, Niels Bohr LIbrary).

proposed by Steven Weinberg and Abdus Salam (fig. 10.9), that unites electromagnetism and the weak force (or weak interaction).

The prototypical weak force process is beta radioactivity, in which a neutron (which can be free or in a nucleus) decays into a proton, an electron, and an antineutrino. That is, the neutron disappears, and three new particles appear in its stead.[11] Thus it is on the surface a four- rather than a three-pronged vertex. Theorists long supposed that this vertex could eventually be analyzed as a combination of two three-pronged vertices, connected by a hypothetical virtual particle, termed the *intermediate vector boson.* Experiments involving weak processes were carried out intensively in the 1960s and 1970s, when it became possible to produce copious supplies of neutrinos in accelerator laboratories. (The neutrino beam is important here because, although many elementary particles participate in the weak interaction—protons, neutrons, electrons, muons, among others—the neutrino is the only one that participates *only* in weak interactions. Other particles participate in electromagnetic and/or strong interactions as well, and these effects easily mask the influence of the weak interaction.) Evidence accumulated that the intermediate vector boson would have a very large mass and come in three varieties (like the pion), positive, negative, and neutral. The large mass is associated with the extreme short range of the interaction (less than 10^{-17} meters) and implies that it can be created only in extremely energetic collisions. Finally in early 1983 the charged intermediate vector bosons were discovered at CERN, the European high-energy research laboratory in Geneva. In agreement with theory, they are about eighty-six times as massive as the proton. A

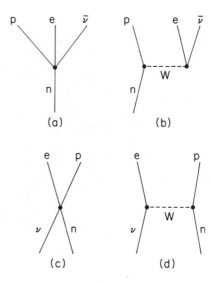

Fig. 10.10. Weak interactions. (*a*) The process of nuclear beta decay according to the older theory, based on a four-pronged vertex. A neutron decays into a proton, an electron, and an antineutrino. (The Greek letter nu [*ν*] is used for the neutrino; *ν* with a bar over it is an antineutrino; Latin *n* represents the neutron.) (*b*) Beta decay as described by a theory based on a three-pronged vertex and the intermediate vector boson (*W*), long hypothesized and recently discovered. A neutron decays into a proton and the boson *W*. Note that the boson must be negatively charged to balance the positive charge created on the proton. The boson later decays into an electron (which then carries the negative charge), and an (uncharged) antineutrino. The exchanged particle has transferred electric charge. Negative, positive, and neutral bosons have all been found. (*c*) A process that can be studied when beams of neutrinos are available in the laboratory. A neutrino and a neutron collide and an electron and a proton emerge. Note that total charge remains zero. This process is very similar to (*a*), the only difference being that instead of the creation of an antineutrino, we have the destruction of a neutrino. (*d*) The same process as (*c*), interpreted in terms of the exchange of an intermediate particle. Notice the similarity to electron-electron scattering (see fig. 10.7*c*) with photon exchange.

few months later the uncharged member of the family was discovered. (see fig. 10.10.)

The Weinberg-Salam theory formulates the electromagnetic and weak interactions as two aspects of a single comprehensive quantum field theory and places the photon together with the three intermediate vector bosons into a family of mediating particles. Although the photon and the vector boson differ so substantially in mass, they do possess certain similarities, notably intrinsic spin of the same magnitude. Moreover, the large mass of the vector boson in a sense explains the fact that the weak interaction is weaker (meaning that the probability for a given process is small) than electromagnetism. The intrinsic strength of the weak interaction is comparable to that of electromagnetism, but it appears weaker because it acts only over a very small region of space.

The successes of the new *electroweak* theory have encouraged theorists who aim at unifying the strong interaction with the electroweak, placing the photon and the vector bosons into a still larger class of exchanged particles. It may be that at extremely high energies, beyond what is available in laboratories today, we will find a coalescence of what today appear to be unrelated forms of interaction. New approaches to the unification of gravity with the other interactions are also being studied. Where these speculations will lead we cannot say. But experience shows that wherever nature takes us, we will find not an end but a beginning.

APPENDIXES

The Concept of a Field

Some quantities may be specified by giving a single number or just a few numbers: someone is 6 feet tall; a box is 3 feet by 4 feet by 5 feet. Other quantities must be specified by giving a large (perhaps infinite) array of numbers. To give the temperature distribution throughout the United States we must attach a number, the local temperature, to every spot on the map. (In practice we only need see a representative sample of temperatures at various places around the country to get a reasonable idea of the weather. But in principle each spot has its own temperature.) This series of numbers is called a *field,* actually a continous distribution of numbers throughout a region of space. Atmospheric pressure is another example of a field. It is clear that both these fields can be extended to three dimensions, giving the temperature and pressure at various heights above the earth's surface. Furthermore fields can vary with time; today's weather map is not the same as yesterday's.

Sometimes the variation of a field over time and space takes the particular form that we call a wave. If at a certain instant the field quantity varies in a periodic way over distance, and if, as time goes on, the periodic pattern moves in some direction at a definite speed, then this field is a wave. One of the great accomplishments of classical physics was to show mathematically how such traveling waves in space might arise. A wave can exist in the pressure of the air, and this is what we call sound. Maxwell's elegant mathematical analysis showed that waves could occur in the electromagnetic field.

In the writings of Michael Faraday the field concept began to emerge as a way of thinking about the interaction of two separated bodies, such as two magnets pulling on each other across a distance or the earth-moon system interacting gravitationally. The prevailing view of interaction, following Newton, was that of *action-at-a-distance*—that two objects interact and exert forces on one another without the intervening space playing any role in the process. Faraday's ideas evolved from his extensive experimentation with electricity and magnetism and in particular with the effects of intervening media. He found that an electric charge produces a certain force on another nearby and that this force remains (although weakened to some degree) if certain solid or liquid matter is placed between the two. He pictured this matter as in some way carrying the force from one charge to the other. An influence exists not only at the charges but in the matter between. Rather than action-at-a-distance, each charge acts only on the matter immediately surrounding it; the influence spreads step by step, from each piece of matter to the piece adjacent to it. The influence that has spread through space is a field, the electric field. Similarly a magnet establishes a magnetic field.

It is a long step, but to Faraday it was the logical step, to extend the field concept to empty space. The spreading of electric influence across air or a vacuum is entirely similar to (though quantitatively different from) that across solid or liquid matter, and so it is natural to picture fields extending in a similar way across empty space.

Blackbodies

Two blackbodies at the same temperature will emit the same radiation. Proof of this statement can be given by the following "thought" experiment, a kind of argument based on a hypothetical experiment, one that need not be performed but that logically implies important general conclusions. The argument here is based on the nature of temperature. If two objects of different temperatures are brought together, energy flows from the hotter object to the colder until, after some time, they reach a common temperature. The hotter object has become cooler; the colder object has become warmer. (Consider, for example, placing a just-boiled egg in a pot of cold water.) But if two objects of the same temperature are brought together, they do not change. They remain in a condition of equilibrium, and we never observe that one object becomes colder, the other warmer. (Consider an egg at room temperature placed in a pot of water also at room temperature.) A broad generalization based upon this kind of phenomenon is known as the second law of thermodynamics.

In the case of two black surfaces (A and B) at the same temperature, suppose one (A) radiates more energy than the other. We imagine placing the two surfaces next to one another and allow each surface to absorb all the radiation emitted by the other (since they are black). Thus B absorbs more radiation than it emits and, receiving a net increase in energy, it becomes hotter. A, correspondingly, becomes cooler. Thus the second law of thermodynamics would be violated; our assumption that A radiated more than B must be false.

A similar argument shows that, for a given temperature, a blackbody is the best radiator. Consider A, a blackbody, and C, which is not a blackbody but absorbs only part of the radiation it receives. Is it possible for C to radiate more than A? Let us bring A and C together. C absorbs not all the radiation but less than all the radiation that was emitted by A. But C, by hypothesis, emits more than was emitted by A. For C, in other words, there is a net loss of energy, and C will cool down. This again violates the laws of thermodynamics.

What makes a hole admitting to a large cavity a good black surface? It is simply that light falling upon the hole will enter the cavity and will have a small chance of emerging. The probability of the light beam hitting the hole from inside is very small, just because the hole is small; and so light will bounce around inside, striking the walls many times, before chancing to hit the hole. Even if the walls are not very dark (i.e., absorptive), as long as they absorb some incident light, almost all the light will be absorbed before any is able to escape the cavity.

De Broglie Waves

The concept of the de Broglie wavelength of a particle can be tied to quantization of the orbits of an electron in the hydrogen atom by analogy with waves on a string. Consider for a moment waves on a taut guitar string, attached at both ends. When the string is plucked, it may oscillate in a number of different modes (fig. C.1). In the simplest mode the entire string (except the endpoints) moves up and down together; or it may move as two separate units, one half moving up while the other moves down; or it may move as three or more separate units. These motions are called *standing waves*. They are closely related to the kind of waves we have discussed in chapter 1 and elsewhere, which we may now call *traveling waves*.

In any wave there is some quantity that varies in a periodic way over space, with alternating crests and troughs, as in figure 1.16. The quantity could be the height of a water surface or of a stretched string, or the strength of an electromagnetic field. As time goes on this entire alternating *pattern* may move in some direction at the wave speed. Then we have a traveling wave. The pattern travels, let us say, along the length of a string, but the bits of string do not travel in that way; they only oscillate up and down (perpendicular to the length of the string).

The motions of the string in figure C.1 are quite different because the string is constrained at the endpoints. Here each point we show as a

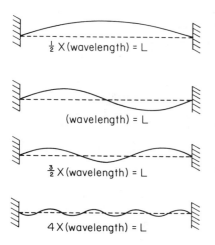

$\frac{1}{2} \times$ (wavelength) = L

(wavelength) = L

$\frac{3}{2} \times$ (wavelength) = L

$4 \times$ (wavelength) = L

Fig. C.1. Some of the possible modes of oscillation of a stretched string. The fact that the string is fixed at both ends places limitations on the possible values of the wavelength. The rule is that one-half the wavelength times an *integer* must equal the length, *L*, of the string. In the cases shown in the diagram the values of the integer are 1, 2, 3, and 8. All integers are possible, but the wavelength is still "quantized."

Fig. C.2. The de Broglie concept of the electron as a wave in the hydrogen atom. The wave must be single-valued as it goes around the closed circle of the electron's "orbit." Therefore there is a limitation on the possible values of the electron's wavelength. In this case an integer times the wavelength must equal the circumference of the circle. This condition quantizes the wavelength; that in turn quantizes the radius of the orbit and the electron's energy. It should be understood that the wavy line does not represent the path of the electron. It only indicates in a schematic way the relation between wave properties and quantization.

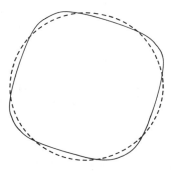 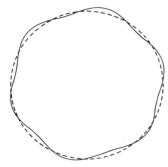

4 X (wavelength) = circumference   6 X (wavelength) = circumference

crest will move, as time advances, down to a trough, up to a crest, down again, and so on. Similarly the troughs move up and down. The points where the string is shown crossing the (dashed) horizontal line do not move; they are called nodes. The effect is that of a string oscillating in a complex way, but there is no apparent motion to the left or right. This is a standing wave. It has a well-defined wavelength and also a well-defined frequency (the rate at which each bit of string oscillates). The same inverse relation between frequency and wavelength holds for these standing waves as holds for traveling waves on the string.

The special property of standing waves is that the wavelength is quantized. The wavelength cannot take on any value but only a discrete set of possible values simply related to the length of the string. Quantization occurs because the string is constrained at both ends. The frequency is, therefore, also quantized. If you try to set up standing waves on a string by holding one end and vibrating it, you find that there is a strong wavelike response only if you employ just the right frequency to match one of the allowed wavelength values. These are called the resonant frequencies.

You can get a feel (literally) for these resonances by using a slinky instead of a stretched string. The frequencies of the guitar string are several hundred oscillations per second and so you cannot match them by hand (although it is not difficult with an electronic oscillator), but the resonant frequencies of the slinky are much lower. Attach one end of the slinky to a hook held firmly in a wall, and stretch the slinky out as far as you can without unraveling any of the coils. Then try to set up standing waves by shaking the end you are holding up and down in small motions. At a fairly slow rate (around one or two oscillations per second) you may hit the resonance of the lowest mode. The slinky will swing vigorously up and down as a unit. Now increase the frequency, and try to find the next mode; it should be at twice the frequency of the lowest mode. Here the slinky oscillates in two units, as in the second example of figure C.1. Finding these resonances is tricky, because the slinky insists that you match its discrete resonances closely, and that is difficult to do in freehand motion. Try hitting the third and fourth mode. Notice that between resonances the up-down motion is weak and irregular.

Similar effects occur in the case of the oscillating air column in a trumpet or other wind instrument. Here the wave is a sound wave, and the frequency is related to the sound's pitch. As a result of quantization, the open trumpet (without valves to lengthen the air column) can play only a certain discrete set of pitches.

Now imagine a standing wave spread along the circular orbit of an electron moving about a hydrogen nucleus. Because the circle is closed, the wave, after proceeding around the circumference of the circle, must match its starting point. That is, the only allowed wavelengths are those such that an integral number of them fit into the circumference (fig. C.2). The wavelength is quantized.

De Broglie's contribution was to suggest what the relation between the wavelength of the electron and its energy should be, a relation analogous to the Einstein relation for the photon, $\varepsilon = h\nu$. Schrödinger developed this idea into a complete program, called wave mechanics, for calculating the energy levels of not only the hydrogen atom but all atoms. It has turned out to be the most productive formulation of quantum theory, applicable in all areas of atomic and subatomic physics.

White Dwarfs, Pulsars, and Neutron Stars

A main-sequence star is held in equilibrium by the balance of two forces, the inward attraction of gravity and the outward pressure generated by immensely hot gases in the center. What maintains equilibrium in the white dwarf, in which the source of energy at the center has been extinguished? The answer to this question grows out of one of the most fundamental discoveries of modern physics, the *exclusion principle,* enunciated in 1935 by Wolfgang Pauli (fig. D.1). The exclusion principle was invoked to explain the orbital motions of electrons in many-electron atoms and thereby the chemical properties of elements. It states that it is impossible for two electrons to be in precisely the same state at the same time. The effect can be thought of as a force, or pressure, that tends to keep electrons apart from one another. It is as if each electron creates around itself a shell of a certain radius, which nothing from outside can penetrate. The exclusion principle also applies to protons and neutrons, although here the effect may be less evident because the shell, or region of exclusion, is smaller. (The size of the region varies *inversely* with the mass of the particle.)

In astronomy the exclusion principle is important when the matter of a star becomes very dense, so that the regions of exclusion of the par-

Fig. D.1. *Left to right:* Paul Dirac, Wolfgang Pauli, and Rudolf Peierls. (Courtesy AIP, Niels Bohr Library)

ticles making up the star almost overlap; and that is what happens in the case of the white dwarf. The electrons in the matter of the star are pressed so closely together that there comes into play an intense outward pressure tending to keep them apart. It is this "electron pressure" that balances the force of gravity and maintains the equilibrium of a white dwarf. It happens at a density about 10^5 times that of ordinary solids and liquids.

But there is a way in which the electron pressure can be overcome, an outlet that in fact must be taken if the pull of gravity is sufficiently strong. The star is primarily hydrogen—protons and electrons. Under conditions of high pressure and density it is possible for a proton and an electron to assimilate and become a single neutron (the positive and negative charges then canceling in the neutron, which has no electrical charge). This step may seem a bit like black magic, but in fact it is a process closely related to a more familiar nuclear phenomenon. *Beta radioactivity,*[1] discovered in 1899, is a process in which a neutron in a nucleus spontaneously decays (it ceases to exist), and a proton and an electron are created. The proton remains in the nucleus, and the electron streaming out is the "beta particle" making up the observed radioactivity. The formation of neutrons in a massive star is very much like the reverse process. It is a nuclear reaction that can go in either direction depending upon the environment in which the particles find themselves. (A similar process is involved in the conversion of a proton to a neutron when hydrogen is burned in a main-sequence star. In all these processes neutrinos or their antiparticles are created.)

When a massive star loses its internal power, its collapse is so violent that electron pressure cannot withstand it. Electrons vanish, and the neutron star begins to form. Collapse is ultimately stopped when the system reaches such an extremely high density that the neutrons themselves are in contact; or, to be more precise, the region of exclusion around the neutrons, which is much smaller than that around electrons, takes effect. There is a repulsion among the neutrons that now balances the force of gravity and establishes an equilibrium at a density of the order of 10^{15} times that of ordinary matter. Not all matter of the star is trapped in this dense core; much of it explodes into surrounding space as a supernova, releasing prodigious quantities of energy.

Recall that ordinary solid matter, as Rutherford learned in his scattering experiments, is made of small nuclei with large spaces between them in which atomic electrons move. The size of a nucleus is roughly 10^{-15} meters, while the size of an atom (which is also the distance between nuclei) is roughly 10^{-10} meters. If we were to compress this matter, "squeeze out" the electrons, and force the nuclei into contact with one another, we would have a system of about the same density as a neutron star. To put it another way, a neutron star may be thought of as a single huge nucleus.

Hubble's Law and the Cosmological Principle

The *cosmological principle* is the hypothesis that every point in the universe is like every other point. Let each galaxy have its own Hubble, sitting behind a powerful telescope, measuring the positions and velocities of the galaxies that he can see. What is meant by saying "every point is like every other point" is that each one of these Hubbles will discover the same astronomical laws, the same laws found by "our" Hubble. (Of course we are here taking the large-scale view of the universe, ignoring "small" inhomogeneities such as the clustering of galaxies and small motions such as the orbiting of galaxies in a cluster.) Though certainly not proven, the hypothesis is one that many scientists are predisposed to believe, on the not altogether irrational basis of simplicity—that the universe, when looked at properly, is simple.[1]

We observe galaxy *A* (fig. E.1), 18 megaparsecs from us, and find that it is receding from us at 1,000 kilometers/second. Our colleague in galaxy *A* gazes through his telescope and sees galaxy *B*, 18 megaparsecs from him (in the same direction). Now if the cosmological principle is correct, galaxy *B* will be receding *from him* at 1,000 km/s. In other words, the speed of *B relative to A* is 1,000 km/s; but it follows that the speed of *B relative to us,* to our galaxy, is 2,000 km/s.[2] And that is Hubble's law. The galaxy farther from us (namely, *B*) is receding from us at a faster rate (2,000 km/s) than is the galaxy nearer us (*A*).

Notice, though, that the same argument would have worked if the galaxies were moving toward us rather than away from us. Or indeed, as a special case, the velocities could have all been zero. The cosmological principle implies that if the universe is expanding, then the velocities of expansion must obey Hubble's law—the more distant galaxies are moving faster, and furthermore the speeds of the galaxies are proportional to their distances. If the universe is contracting, then the speeds are all toward us, but they are again proportional to distance.

Fig. E.1. The cosmological principle. (*a*) For an expanding universe, an observer in galaxy *A* sees galaxy *B* receding at 1,000 km/s relative to *A*. Hence *B* is receding at 2,000 km/s from the Milky Way galaxy. (*b*) For a contracting universe, velocities are reversed, and a modified Hubble's law holds: galaxies farther away are moving toward us at proportionally greater speeds.

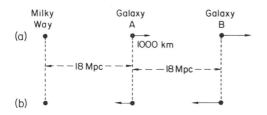

To the extent that Hubble's law is true, two things are implied: (1) the cosmological principle holds, and (2) the universe is expanding. If it had been found that there was *no* proportionality between velocity and distance, we would have been forced to conclude either that the cosmological principle was false or that our observations simply did not extend far enough to detect the large-scale structure of the universe. We should bear in mind, in this connection, that astronomers do not have independent confirmation of Hubble's law at the greatest distances, because distance measurement becomes unreliable.

Fluorescence

F

A graphic view of fluorescence and the relation between electron and vibrational levels can be given by a plot of energy versus interatomic distance (fig. F.1).

For a molecule in its ground state there is a fixed distance between any pair of atoms. For simplicity consider a diatomic molecule, with R_0 representing the distance between the two atoms and E_0 the energy of the lowest electron state. Suppose we could arbitrarily change the spacing between the atoms to some other value, R, greater or lesser than R_0. Then the molecule would have an energy, E, greater than E_0. A graph of E versus R (the solid curve in fig. F.1) shows a minimum at $R = R_0$, since the ground state is by definition the state of lowest energy. It is the state to which the system naturally tends. A good analogy to the behavior of the molecule is given by imagining a vertical track shaped like this graph, with a marble rolling inside it. The marble tends to settle at the point R_0. But if the marble is displaced slightly to the right of R_0 it will roll back and forth executing oscillatory motion about R_0. This motion, between R_1 and R_2 in the figure, is analogous to a vibrational level of the molecule. One important difference between the marble and the molecule is that the levels in the molecular motion are quantized, represented by a discrete series of closely spaced horizontal lines.

Now we consider the molecule in a different electron configuration, that is, with its electrons in different orbits—corresponding to its

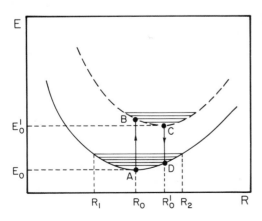

Fig. F.1. Fluorescence

229

first excited state. There is similarly an equilibrium spacing, R_0', between the two atoms for this case; and if we change the spacing to one greater or lesser than R_0', the molecule will oscillate about this new equilibrium value. The dashed curve represents the energy of the molecule in its first excited electron state, plotted as a function of spacing R. The minimum energy, E_0', is greater than E_0 because this is a higher energy state; and R_0' is greater than R_0 because in the excited state the molecule is less tightly bound, and so the two atoms are not held so closely together. Vibrational levels are also shown for the excited electron state.

Suppose the molecule is in its ground state (point A) and receives a sudden boost in energy by absorption of a photon. This event is represented by the vertical arrow, taking the molecule, not to the lowest point in the excited electron state, but to one of the higher vibrational levels of that state (point B). (The assumption here is that during the time that the photon is being absorbed the spacing between the atoms does not change. This assumption is justified by the fact that the nuclei of the two atoms are relatively very heavy, indeed thousands of times heavier than the electrons. And so the time needed for an adjustment of the spacing between the nuclei is much longer than the time needed for an electromagnetic adjustment of the electron orbits.)

The molecule quickly loses its vibrational energy, decaying from point B to point C by emission of a number of far-infrared photons or via other low-energy processes. Then fluorescence occurs, represented by the vertical line from point C to point D. Again the emission of a photon is so fast that the spacing between the atoms does not change, and the molecule finds itself in one of the vibrational levels of the ground electron state. Finally this vibrational energy is lost in the form of heat, as the system decays from point D to A again.

Semiconductors and Photocells

Crystalline solids are normally divided into four classes according to their ability to conduct electric currents: insulators, semiconductors, metals, and the rather more esoteric semimetals. Electrical conductivity is important as a defining property because it is a measure of how the electrons of the assembled atoms are distributed in the solid.

Imagine now a large collection of atoms, initially very far apart, and gradually brought altogether. At first the possible states for a given electron correspond to the ordinary discrete energy levels that give rise to the bright line spectrum of the gas—the ground state, the first excited state, the second excited state, etc., separated by energy gaps. As the atoms are brought closer together there are interactions among them that begin to distort the energy levels; each level is spread into a continuum, forming what is called a "band" of possible energies. Between two bands their remains (or may remain for the upper bands) an *energy gap*. It is the gap and the way the possible energy states in the upper bands are filled (by electrons) that determine the electrical properties of the solid.[1]

In the separated atoms there will be some electrons that are, as discussed in the previous chapter, nonbinding; they are held tightly to the atom and do not interact significantly with neighboring atoms. We may say that these electrons lie on the lowest of the atomic levels. The electrons that do form bonds are those we think of as "outer" electrons, occupying the high energy levels. The atom has numerous still higher levels, which are normally not "occupied" by electrons; they are *potential* spots to which electrons might be raised if they received an appropriate energy boost. When the atoms are brought together to form the crystal, the nonbinding electrons fill the lowest bands, and the outer electrons lie in the highest occupied band. In carbon, for example, two of its six electrons are nonbinding; the remaining four fall into the upper band, which is called the *valence band*. When carbon has the structure of diamond (one of its several variant crystalline forms), the number of states in the valence band is just equal to four times the number of atoms in the solid, and the band is said to be filled with electrons. This means that, like sardines in a can, no electron can "do anything"; it cannot change its state in any way, *unless*—and this may require a substantial input of energy—it is excited to the next higher band. It is the Pauli exclusion principle again that lies behind this argument: an electron cannot move to another state in the valence band, since these states are already occupied, and the exclusion principle forbids two electrons in the same state.

Insulators and semiconductors are materials in which the valence band is filled, all levels are occupied, and there is a gap between it and the

Fig. G.1. Energy levels of a crystalline solid. The cross-hatched areas are energy bands, possible energies for the electrons of the solid. The dots represent electrons. (*a*) A metal; (*b*) an insulator; (*c*) a semiconductor.

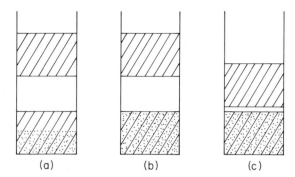

(a)   (b)   (c)

next higher band, which is called the *conduction band* (fig. G.1). Without the input of some extra energy to bridge that gap, no electron can move; there is no flow of electricity. The insulator has a relatively large gap; the semiconductor, a relatively small gap. At normal temperatures the semiconductor can conduct electricity very weakly, a result of thermal excitation of a few electrons. The elements silicon and germanium, which have an electronic structure like carbon (with four outer electrons) and a crystal structure similar to diamond, are industrially important semiconductors.

A metal, on the other hand, is typically built of atoms in which all but one or two of the electrons are noninteracting. The outer one or two electrons, as a result, do not fill the upper band. They are able to conduct electricity, therefore, by simply moving into the next state in the same band; no boost of energy is needed to cross a gap. They are conductors even at low temperature.[2]

Picture the silicon crystal as a system in which each silicon atom is bonded to four neighboring atoms, forming a series of interlocking tetrahedrons; the four outer electrons make up the "glue" that holds the solid rigidly in place. The bond between two adjacent atoms has something of the character of the single covalent bond that acts in an organic molecule. (Because the crystal is a collection of identical geometrical figures, the probability distribution of *each* electron must be thought of as spread throughout the entire crystal. But the distribution is concentrated along the lines between neighboring atoms and still provides the glue of the binding.)

Now add impurities. Suppose we heat a pure silicon crystal in an atmosphere containing the vapor of phosphorus. A small number of phosphorus atoms diffuse into the crystal and finally take the place of silicon at some of the atomic sites of the lattice. Of the five outer electrons of phosphorus, four serve the same function served everywhere else by the four silicon electrons—they bind the crystal in its tetrahedral structure. The fifth electron remains bound, but very weakly, to that atom. It takes only a weak perturbation to remove it from that site and allow it to wander about the material. When this happens, because the solid as a whole is electrically neutral, a net positive charge is left (fixed) at the site. This is the doped *n*-type semiconductor.

The *p*-type semiconductor is produced by doping with an atom that has three outer electrons. At that site there is one fewer electron than there would normally be, and this missing part of the crystal system—a "hole"—is relatively free to move from site to site in the material. It behaves like a positive charge, analogous to the electron in the n-type semiconductor. And when the hole moves, a fixed negatively charged atomic site remains.

Used in a normal electric circuit, the junction between a p-type and an n-type solid has the effect of aiding current flow in one direction and impeding it in the other direction. It thus acts like a diode, the simplest vacuum tube. More complex combinations of doped semiconductors make up transistors, which thus have replaced vacuum tubes as devices to control electric currents in circuits.

In the photovoltaic cell a photon is absorbed and excites a new electron into the conduction band. Far from the junction such an electron would immediately de-excite, falling back into the lower band (either to the state from which it came or to some other hole formed by another photon or by heat) and reemitting a photon. But the junction is designed to capture the electron before it can de-excite and to sweep it away by means of the electric field. Electrons are swept from the p- to the n-side, and holes from the n- to the p-side. An electron arriving at the n-side finds fewer holes; thus it is more difficult for it to decay. With a continuous flux of photons a continuous current of electrons is established across the junction. This is the transformation of light to electrical energy.

Stimulated Emission

Here is Einstein's argument for the existence of stimulated emission. Consider a gas of atoms with two energy levels, E_A and E_B, in a cavity in which there also exists radiation of frequency, ν, where $h\nu = E_B - E_A$. Let X be the rate at which absorption of radiation occurs, and let Y be the rate at which spontaneous emission occurs. Such a system will, after sufficient time, come to equilibrium, a situation in which nothing changes. The *net* number of photons in the cavity and the number of atoms in each state will all be constant. Now if the net number of photons does not change, then the rate at which they are produced must equal the rate at which they are removed: $X = Y$, assuming that spontaneous emission is the only source of new photons. (This equilibrium, we can see, is a dynamic situation—as equilibrium on the atomic level always is; things happen, photons are created and destroyed, atoms jump up and down between levels, but the net [or average] properties of the system do not change.) But, and this is the crux of the argument, X and Y are two quite different things. The absorption rate, X, must be proportional to the number of photons in the cavity: the more photons, the more likely the absorption of one. For example, if we raised the temperature of the cavity there would be more photons (as shown by the theory of blackbody radiation); and so X would increase rapidly at high temperatures. On the other hand, Y, the rate for spontaneous emission, would not depend on temperature at all. Spontaneous emission has nothing to do with the atom's environment; an atom decays in response to its own internal character. It follows that it is impossible to maintain $X = Y$ at all temperatures.

The conclusion is that there must be some third quantity, Z, that is to be added to Y—a second device for producing photons. Z must have the property that, like X, it increases rapidly at high temperatures. More precisely, it must be proportional to the number of photons. And further, it must be the case that at high temperatures both X and Z become much more important than Y. In other words, at high temperatures the important contributions to equilibrium are the excitation of atoms from state A to B, with accompanying absorption of photons (the process X), and the de-excitation from B to A, via the process Z, with accompanying emission of photons.

If we step back and look at what the process Z is, we see that since it is proportional to the number of photons and leads to emission of more photons, we may think of it as emission caused by—or "stimulated" by—the photons in the environment.

Measurements of the
Speed of Light

How can I measure something that travels at 300,000 kilometers per second? Suppose you and I synchronize watches, and you march off across a valley to a hill 1 kilometer away (about three-fifths of a mile). Then at a prearranged time, say it is precisely 1 A.M., I turn on my flashlight, and you try to record the instant that you see the light. The light reaches you $\frac{1}{300,000}$ of a second after 1 A.M. Can you make such a measurement? It seems clear that no matter how accurate the watches, no matter how carefully synchronized, your human reaction time will simply be too slow for you to distinguish the arrival time from exactly 1 A.M.

Galileo tried a similar experiment in the seventeenth century, except that he used a round-trip light path, eliminating the need for synchronized watches. An assistant waited in the dark night for the signal from Galileo's lantern; when he saw it he sent his own light signal back; and Galileo, receiving the return signal, tried to measure the time interval for the two-way trip. Again, the speed of light is too great for a measurement based directly on reactions of the human eye and hand.

But if we look at phenomena occurring over much larger distances, on an astronomical scale for example, we will find light travel times substantially larger—large enough to be measured by straightforward means. It takes a little over 8 minutes for light to travel from the sun to the earth. Not surprisingly then, the first successful determination of the speed of light was based upon observation of motions in the solar system. The planet Jupiter has numerous satellites, several of them large enough to be easily seen with a small telescope. Each of these satellites is regularly eclipsed when it moves behind the giant planet in the course of its orbit; and the timing of these eclipses was being studied by the Danish astronomer Ole Roemer around the year 1676. Roemer found that these eclipses, occurring about every 42 hours, were early by several minutes at times of the year when the earth was close to Jupiter, and late at times when the earth was farther from Jupiter. Roemer correctly concluded that the eclipses occurred regularly and that the time variation was due to the variation in the travel time of light from Jupiter to earth. A calculation based on Roemer's observations and solar system data as known in the seventeenth century gives a speed of about 200,000 km/s.

About a half a century later, Bradley's observation of the aberration of starlight provided a second and more accurate astronomical determination of the speed of light. For the angular shift in the position of a star depends on the ratio of the speed of the earth in its orbit to the speed of light.

In 1849 the Frenchman Fizeau (fig. I.1) carried out the first ter-restrial measurement of the speed of light. As Galileo had done, Fizeau attempted to time a round-trip passage of a light beam. But instead of having the signal returned by a human collaborator, Fizeau reflected his beam from a mirror. The mirror was at a distance of around 8.6 kilo-meters—a round-trip path of 17.2 kilometers—a time of roughly $\frac{1}{17,000}$ of a second. To measure such a minute time interval, Fizeau used a trick characterizing many clever high-precision experiments. His ex-perimental arrangement (fig. I.2) essentially multiplies the time interval by a large (and known) factor, and the resulting long time interval is mea-sured. He built a rotating wheel and around the circumference cut 720 teeth separated by 720 gaps. Light strikes the edge of the wheel (perpen-dicular to the face) as the wheel rotates. If there is a tooth at that instant the beam is blocked; if there is a gap the beam passes through. Now sup-pose the time it takes for the wheel to turn from a gap to the adjacent tooth is $\frac{1}{17,000}$ of a second. Then light that leaves the source and passes through a gap will travel 8.6 kilometers to the mirror, return to the wheel's edge, and get there just in time to meet the oncoming adjacent tooth. The light is blocked. In other words, if the rate of rotation of the wheel is at just the right value, the viewer, looking for the reflected light signal, sees no light. Thus Fizeau has replaced the task of recording the instant of light's arrival with the task of observing darkness or light.

How fast must the wheel rotate to achieve this null condition? Since there are 720 teeth and 720 gaps, the time we are talking about is $\frac{1}{1,440}$ of the time for a complete revolution of the wheel. Suppose the wheel makes a revolution in $\frac{1}{10}$ of a second. Then the time from gap to tooth is about $\frac{1}{14,000}$ of a second. This is almost fast enough. Speeding up the wheel slightly brings the time down to $\frac{1}{17,000}$ of a second, and the condition of darkness. Thus Fizeau has multiplied his time interval by 1,440; he is left with the relatively simple task of measuring the rate of rotation of a wheel turning ten to fifteen times per second.

A modification of Fizeau's experiment was shortly introduced by Foucault, who replaced the toothed wheel with a rotating many-faceted mirror. One face of the mirror sends the light beam on its way to the dis-tant reflecting mirror, and the returning beam is caught by the next face. This technique, more accurate than Fizeau's, provided a value of the speed of light good to three significant figures. The experiment was re-peated in 1883 by the American physicist Albert Michelson, who con-tinued to refine and improve the experiment over the course of fifty years. Michelson's last experiment, completed shortly after his death in 1931, gave the value 299,774 km/s, with an uncertainty of 11 kilometers. (Later measurements, as we shall see, indicate that this uncertainty was slightly underestimated.)

While the obvious method of measuring speed is to take distance divided by time, an alternate approach presents itself in the case of a wave, for the wave's speed is the product of its frequency and its wave-length. Thus independent measurements of these two properties can be used to determine the speed. Through most of the history of the wave theory of light, only wavelengths were measured. Frequencies in the re-gion of 10^{14} hertz were too large to be measured directly and were deter-mined from the wavelength and the known speed of light. But once Maxwell and Hertz had established that light was an electromagnetic wave and that all electromagnetic waves traveled with the same speed, it became possible to measure the wave speed by using waves outside the visible range. Indeed, Hertz himself measured the wavelength and the frequency of his first radio waves and determined that their speed was the same as that of light.

In 1950 this approach was employed with microwaves in two experiments done independently in England and the United States. Microwaves are electronically produced electromagnetic signals, similar to radio signals used for broadcasting but of considerably shorter wavelength (hence the prefix *micro)*. Microwave technology was developed into a highly sophisticated science in the years during and after World War II, because of its importance in radar. The experiments on the speed of light made use of microwaves confined to a cylindrical cavity known as a *cavity resonator*. In effect these were standing waves, analogous to the standing waves that may be set up on a stretched string of wire. (see app. C.) As in the latter case, standing electromagnetic waves can be set up only at certain frequencies, the resonant frequencies of the cylinder. In the cavity resonator an electronic oscillator sets up electromagnetic waves at these resonant frequencies. The frequency of the waves is known, and the wavelength is related in a simple way to the length and diameter of the tube (just as the wavelength on a string is related to the length of the string). Results of the two experiments were the values 299,792.5 and 299,789.3 km/s, with some uncertainty in the sixth digit.

Two years later an experiment with free microwaves (waves not confined to the interior of a cylinder) gave the result 299,792.6 km/s. Here the wavelength was determined in an interferometer, a device similar to Michelson's interferometer, which splits the light beam into two parts and looks for interference between the two.

This brief discussion of these high-precision measurements passes over a host of intricate technical problems and ingenious solutions. Microwave frequencies are in the region of 10^{10} hertz, and it is a considerable feat to measure anything varying that rapidly. One technique is based on generation of what are called *harmonics*. In the case of a sound source—a piano string, for example—an oscillation at a certain frequency, *f,* is usually accompanied by additional components at frequencies twice *f,* three times *f,* four times *f,* and so on. These are called harmonics, and they determine the quality of a sound; they make the note middle C sound differently on the clarinet than on the violin. In the case

Fig. I.1. Armand-Hippolyte-Louis Fizeau (Courtesy AIP, Niels Bohr Library, E. Scott Barr Collection)

Fig. I.2. Fizeau's device for measuring the speed of light. A lens focuses light to a small point at the edge of the wheel, *W.* Light is then either blocked by a tooth of the wheel or passes through a gap. Light travels a distance of 8.6 kilometers to the mirror, *M,* and back to the rotating wheel. The half-silvered plate, *P,* at 45 degrees, passes some light through from the first lens to the wheel. It also reflects some of the returning signal downward to the viewer.

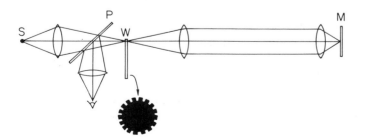

of an oscillating electrical current, its harmonics can be produced by passing the current into a nonlinear circuit element, that is, a device in which the voltage is not precisely proportional to the imposed current. By generating harmonics in one or more stages from precisely known low frequencies, physicists can create very precisely known high frequencies (since the numbers that multiply frequency are exact integers).

Now to measure an unknown high-frequency oscillation, we look for a known high harmonic that will be very close in frequency to the oscillation to be measured. Then we make use of another property of oscillatory systems (also familiar in the case of sound), namely, *beats*. When two sounds very close in frequency are heard together, the result is a pulsating sound of one pitch, a tone that slowly and rhythmically alternates between loud and soft. This effect is called beats, and the beat frequency, the rate of pulsation, is equal to the difference between the two original frequencies. (For example, if two guitar strings are tuned to 440 hertz and 442 hertz, then when they are sounded together there are two beats per second. Guitar players and piano tuners use beats for tuning strings.) In the case of electromagnetic oscillations, a known and an unknown high-frequency signal, both perhaps in the region of 10^{10} hertz or more, can be combined to produce a beat frequency that may be as low as 10^8 hertz. The beat frequency is now low enough (it corresponds to a time of a hundred millionth of a second for one oscillation) to be measured fairly easily with standard electronic techniques. From the beat frequency and the known high frequency, the physicist can calculate the unknown high frequency.

Scientists studying electromagnetic waves find themselves in a box between frequency and wavelength. As they go to high frequencies it becomes more and more difficult to make accurate measurements because each oscillation takes place in an extremely short time. But lower frequencies mean longer wavelengths, and, paradoxically, it is more difficult to measure long wavelengths than short. Wavelengths are measured by looking for the interference of two waves that have traveled different path lengths. We have seen this in Young's two-slit experiment, in Newton's rings, and in the interferometer. The length of a single wave is determined in terms of a physical distance in the apparatus, a distance along which the light ray travels. But the concept of a light ray's "path length" is only meaningful in a situation in which the spreading of light (i.e., diffraction) can be neglected. That condition, when the ray concept is meaningful, holds well only for the case of short wavelengths. Interference experiments give very accurate wavelength values for radiation in the visible part of the spectrum, less accurate in the far-infrared, and still less accurate in the microwave region.

Wavelength determinations in the visible have always been among the most accurate of all physical measurements. Indeed, in 1960 scientists agreed to use the wavelength of a particular atomic transition in the element krypton as the international standard of length. The meter was *defined* as a particular number times the wavelength of this spectral line (as opposed to measuring the wavelength in terms of a physical meter stick). All other length standards (such as meter sticks) were then measured in terms of the krypton line.

Scientists who wished to measure the speed of electromagnetic radiation by measuring both frequency and wavelength therefore sought to move up from microwaves to visible light, where wavelengths can be measured to nine significant figures. The goal was to make direct measurements of the frequencies of visible light, which previously had been determined only by calculation from the speed of light. This was accomplished in the early 1970s by several groups, using techniques like those described above—starting with lower-frequency oscillators, generating harmonics, using beat frequencies to measure higher frequencies, and moving up in several stages from the microwave to the infrared to the visible. These experiments would not have been possible without a variety of sophisticated modern tools, including a series of lasers in different spectral regions and techniques to maintain very stable laser frequencies. These tools have now been developed to the point where the accuracy of frequency measurements is substantially greater than that of wavelength measurements. Values for the speed of light found in two independent experiments are 299,792,457 and 299,792,459 *meters* per second, with uncertainty lying only in the last figure.

These results are so precise that they exceed the precision of the krypton length standard—that is, they exceed the degree of reproducibility of measurements based on this standard. Therefore a new length standard was adopted in 1983, based upon the speed of light. The international meter is now defined as the fraction $1/299{,}792{,}458$ of the distance traveled by light in 1 second. (The international time standard—that is, the definition of the second—is itself an atomic standard, based on the frequency of a transition of a cesium atom. Earlier time standards were astronomical, based on the time of the earth's rotation on its axis.) Thus the speed of light, given in meters per second, is now a defined quantity, being exactly 299,792,458 m/s.[1]

Although it is no longer possible to "measure" the speed of light in meters per second (because the meter is now defined in terms of the speed of light), it is possible to seek higher and higher accuracy in frequency and wavelength measurements. In the near future optical techniques may be precise enough to measure frequencies to fourteen significant figures. This will mean that, for the first time, we will be able to count the exact number of light wave oscillations in a second.

The new status of the speed of light does not imply that it is, in some sense, no longer an important physical quantity. It merely emphasizes that the meter and the second are unimportant quantities; they are arbitrary values chosen for convenience by today's scientists. From the standpoint of nature, the value of the speed of light relative to the meter and the second is of no significance. But the significance of any other velocity (say, the speed of an electron in its atomic orbit or the velocity of matter in an exploding galaxy) lies in its value relative to the speed of light. The new length standard recognizes this fact explicitly.

The Strong Interaction, Pions, and Quarks

J

Suppose the mechanism of a force is believed to be the exchange of pions. In the emission of a virtual pion, energy conservation must be violated, and the extent of this violation must be at least the energy associated with the pion's rest mass. The time during which the virtual particle exists is inversely proportional to the extent to which energy conservation is violated. Thus the lifetime of a virtual particle with rest mass is shorter than that of a virtual particle with zero rest mass. When a virtual particle has a shorter lifetime, it travels a shorter distance, and therefore the range of the force which it mediates will be shorter. The massless photon creates the electromagnetic force, which has an essentially infinite range (although the force becomes weaker as the inverse square of the distance between the two interacting particles); the pion creates a short-range force. In general the range of a force is inversely proportional to the rest mass of the particle mediating it. Yukawa worked backward from the known range of nuclear forces to predict the rest mass of the hypothetical mediating particle.

The quantum field theory of strong interactions is not as well established as quantum electrodynamics. The primitive vertex here is the emission of a pion by a nucleon (the generic term for a proton or neutron). Because the interaction is much stronger (when within the appropriate range) than electromagnetism, any given process involves much more complicated combinations built out of this vertex, and so calculations based on the theory become much more difficult. Furthermore, developments over the last two decades have uncovered a deeper level in the structure of matter, a level out of which the world of pions and nucleons is to be constructed. Here are the particles called *quarks,* the subunits of both pions and nucleons. Interactions among quarks are also strong, and a new set of exchanged particles, dubbed *gluons,* has been posited—the "glue" that holds quarks together. The fundamental three-pronged vertex, inherited from the theory of light, remains. But now it takes the form of a quark emitting a gluon.

Notes

Chapter 1

1. Naming of the colors of the spectrum is somewhat arbitrary since there is really a continuum of color. Newton distinguished seven colors: red, orange, yellow, green, blue, indigo, and violet. Some observers listed fewer colors; others listed more.

2. Sir Isaac Newton, *Opticks* (New York: Dover Publications, 1979), 20 and 26.

3. These effects should not be confused with the more commonly observed *penumbra,* or partial shadow, at the edge of a shadow. The penumbra is a consequence of the fact that the light source is larger than a point. Some beams from the source are blocked by the obstacle, other beams are not, and a partial shadow is produced.

4. Grimaldi also preceded Newton in the combination of colored lights to produce white light, although Newton did not refer to this work in his *Opticks.*

5. Here we may make an analogy to the passage through the earth of seismic waves caused by motions in the earth's crust. Such waves are both transverse and longitudinal. The longitudinal component passes through solid and liquid; the transverse component passes only through solid. By studying these waves geologists have learned that there is a liquid core near the center of the earth, believed to be molten iron. Seismic waves involve compression and twisting of matter in the body of the earth.

6. James Clerk Maxwell, "A Dynamical Theory of the Electromagnetic Field." See *The Scientific Papers of James Clerk Maxwell* (New York: Dover Publications, 1965), 1:535.

7. A nanometer is one-billionth of a meter. It is a convenient unit for discussing light wavelengths and also for describing atoms and molecules. The abbreviation for it is nm.

8. The unit oscillations per second is now called the *hertz* (Hz). Frequencies used for modern broadcasting are measured in megahertz (one million hertz) or kilohertz (one thousand hertz).

Chapter 2

1. A convex lens is thicker at the center than at the edges. A concave lens is thinner at the center than at the edges.

2. Remember, the rays of light do not actually pass through the virtual image but only appear to diverge from it; so the final image need

not be located inside the microscope tube but can be in the table under the instrument, under the table, or even infinitely far away.

3. For many surfaces in everyday life this analysis is over-simplified. Light may penetrate into a thin layer beneath a surface, be partially absorbed, scattered, reflected many times from small particles (of paint, for example), and finally emerge into the air in all possible directions.

4. Modern techniques for research on visual cells are discussed in the next chapter. Here we simply note that one can fairly easily extract rhodopsin from the cells and study its optical properties in the laboratory.

5. Try this experiment. Stand in a closet with the door opened onto a brightly lit room. Hold a book far from your eye but near enough so that you can read the print. Now gradually close the door until the light becomes dim. There is enough light that you can distinguish the dark of the print from the white of the page, but you cannot read the letters. The acuity of your rods is too poor.

6. One thing we know is that the rhodopsin molecule is a dipolar structure: it has an axis. In the visual cell of some insects rhodopsin molecules have their axes aligned, a property that allows them to detect the polarization of light. Some types of bees and ants navigate by observing the polarization of the light of the sky.

Chapter 3

1. The experiments discussed here involve the mixing of different colored lights, not the mixing of pigments. Mixing blue and yellow paint, for example, will not produce white paint. This is another subject, to be discussed later.

2. There is one exception. If one of the three choices for primaries can be matched by a combination of the other two, then we would have effectively chosen just two primaries. In this case the three chosen colors would not suffice to match all others.

3. The situation is analogous to the three dimensions required to locate an object in space. These may be three coordinates (x, y, and z) that specify the object's position with respect to three perpendicular axes. Or we may choose as one dimension the distance to the object from a fixed origin point. This distance of course does not locate the object, because in addition we have to specify a *direction* in space—an arrow pointing toward the object. To specify direction we must give two angles (for example, one up-down angle and one side-to-side angle), leading to a total of three dimensions. The number of dimensions is always three, although there may be different ways to define these dimensions.

4. The study of rhodopsin has been a much easier task (and was done earlier), because there is more rod pigment, and it is more readily extractable.

5. Cut a slice of rare roast beef under the fluorescent light in the kitchen. You will think it is well-done. Now hold it under an incandescent light. It is rare again.

6. This is not to imply that either the spectrally pure or the spec-

trally broad yellow is highly saturated. Saturation is a psychophysical variable, often but not always correlated with spectral purity in the stimulus. We have just seen that in the yellow region saturation is fairly independent of spectral purity. And for human subjects these yellows are seen as less saturated than spectrally pure light in the green or blue; yellow, in other words, is closer to white.

7. A more complicated situation occurs when we view a complex scene with many colored areas. The eye constantly darts about the scene, concentrating first on one area and then on another, so that there is both successive and simultaneous contrast.

Chapter 4

1. Why not blue? And why is the sun not blue? As we have seen in the previous chapter, a given wavelength distribution has a certain perceptual effect, which cannot be predicted by physics alone. Sunlight is white. A distribution of shorter wavelengths (as from the sky) appears blue.

2. To be precise, we are comparing the energy emitted per second from a fixed unit area (say, a square centimeter) of the surface.

3. Not to be confused with the alternating current induced by Hertz's radio waves, the photoelectric current is a one-way flow of charge from an illuminated metal surface to a second surface nearby.

4. One may ask, why could two quanta not strike the same electron at the same time, thereby supplying double the energy. Further analysis shows that the probability of this happening, with the light intensities normally available, is extremely small.

5. J. Needham and W. Pagel, eds., *Background to Modern Science* (New York: Macmillan, 1938), 68.

6. If you flip a coin 4,000 times, the probability is extremely small, although not zero, that you will produce heads 3,500 times. Similarly, if you think of the alpha particle as being scattered slightly either left or right in each collision with a single atom, it is very unlikely that the number of left scatterings would greatly exceed the number of right scatterings. Hence the probability that a large number of small scatterings will add up to a total large-angle deflection is extremely small.

7. To pursue the matter a bit further, although h is fixed, we can also imagine using photons of very low frequency; the photon's energy, $h\nu$, becomes smaller, and its disturbance of the system being measured is minimized. But low frequency means long wavelength. And as soon as the wavelength becomes long, the *resolving power* of the optical instrument (e.g., a microscope) becomes poor. We cannot obtain information about an object smaller than that wavelength; we cannot even locate the object, except with an uncertainty of the same dimension as that wavelength.

Chapter 5

1. On a more subtle level, it can be argued that if air molecules were distributed precisely uniformly in space, then scattered waves

from nearby molecules would cancel (destructive interference), and there would be no skylight. Rayleigh showed that random molecular motions (associated with the temperature of the air) would remove this cancellation.

2. Atomic oxygen is a gas of single oxygen atoms; it is found in the upper atmosphere. Its properties (including its spectrum) are very different from those of ordinary molecular oxygen.

3. When the sun is out it is impossible to detect the airglow against the bright background of the sky. But from rockets several hundred miles up, above most of the atmosphere, instruments can look down and detect the day airglow.

Chapter 6

1. In fact, the position of a star in the sky does change, but because stars are so far away these changes are very slow (just as an airplane traveling high in the sky at 500 mph appears to move slowly). This real motion is too small to be seen with the naked eye. But with a telescope it can be detected for many of the nearer stars, in observations over a few years or decades.

2. A *light-year,* the distance that light travels in one year, is equal to 0.31 parsecs.

3. A star does not look black because it generates its own light. But it absorbs most of the radiation that strikes it from outside. Therefore it is a blackbody, unlike the moon and earth, which are rather reflective.

4. Temperature correlates with a star's predominant color and also with the types of dark lines in its spectrum (see fig. 6.2). The Hertzsprung-Russell diagram was based on a stellar classification according to spectral lines.

5. Recent experiments have raised some doubts about neutrinos traveling at the speed of light. At present these observations are considered inconclusive.

6. Small internal energy sources exist in Jupiter and Saturn. This energy, however, is not the product of nuclear burning but is believed to be left from the original gravitational collapse of the planet.

7. For a while it was thought, not entirely seriously, that signals with such extreme regularity could have been sent by an intelligent civilization. This idea was quickly dropped when it was realized that the pulsar's signal comes in with a whole mixture of frequencies, whereas a technological society would undoubtedly use a single frequency for most efficient transmission, as our radio stations do.

8. The limit of nuclear burning of a massive star is a result of the properties of nuclei. In fusing hydrogen (1 proton), helium (2 protons, 2 neutrons), carbon (6 protons, 6 neutrons), and other light nuclei, energy is released because the mass of the fusion product is less than that of its constituents. The mass difference is released in the form of energy. This is the case for fusion up to the nucleus of *iron* (26 protons, 30 neutrons). Fusion of nuclei heavier than iron would require an input of energy, rather than releasing energy. Thus there is a natural limit to nuclear burning, even in a massive star. At this point it can only collapse.

Chapter 7

1. How does a hydrogen atom get excited to the upper spin level, so that it can subsequently decay to the lower level? The energy difference between the states is so small that the energy needed for excitation can be drawn from the energy in the random motions of the atoms when two collide. This collision energy is not nearly enough to excite the atom to its first excited *orbital* level, and so no visible quanta will be emitted.

2. This conclusion about the size of the quasar follows from the observation that quasar brightnesses vary on a time scale of several weeks or months. Because of the finite velocity of light, if a quasar were larger than, say, 10 parsecs (about 30 light-years), then light from the more distant parts of the quasar would reach us several years after light from the nearest parts. If the source possessed a pattern of time variation, we would receive, not that single pattern, but a whole series of patterns, displaced in time by amounts from zero to several years; that is, we would receive a muddled signal. It would be like hearing a song sung by a chorus in which one singer starts alone, a second singer begins the song one second later, a third singer one second after that, and so forth. If we *do* receive a clear signal from the quasar, and its time variation is less than a year, it must be that the dimensions of the quasar are considerably less than a light-year. Then the time displacements (of the signals from different parts of the quasar) will be small. This ingenious argument is also used in the case of the pulsar, where the time variation of the order of a second implies that the source must be smaller than one light-second (the distance light travels in a second).

3. Some astronomers have questioned the cosmological interpretation of the redshift; that is, they have questioned whether a large redshift necessarily implies an object at great distance. This doubt has arisen from the observation of certain pairs of objects that are close in the sky, have quite different redshifts, and appear to have a faint bridge of luminosity connecting them. Of course, if there were luminous matter connecting them, it would follow that they were close together in three-dimensional space and hence that the redshift values could not be a correct indicator of distance. Although some controversy has been aroused, most astronomers find the evidence doubtful and feel that the standard interpretation of the redshift remains generally valid.

4. The phenomenon of temperature, which we normally define in terms of the melting point of ice and the boiling point of water, has a much deeper meaning related to the properties of molecules (or atoms or subatomic particles). Specifically, temperature is a measure of the *average* energy of molecules. As matter approaches a state in which molecular motion ceases, the temperature approaches a particular value called "absolute zero." This temperature is about minus 273 degrees on the Celsius scale.

Chapter 8

1. *Charles Darwin's Diary of the Voyage of the H.M.S. "Beagle"*, ed. Nora Barlow (Cambridge: Cambridge University Press, 1933), 107.

2. This phenomenon is not phosphorescence according to the modern definition of that term, although originally the two kinds of light were classified together. (We return to fluorescence and phosphorescence later in this chapter.) The element phosphorus was given its name from the phenomenon, which was known to occur in various materials. The phenomenon of fluorescence, on the other hand, was given its name from the mineral fluorspar, a fluorine compound that fluoresces.

3. *Luciferin* and *luciferase* are generic terms, referring to similar types of substances in different bioluminescent species. The chemical formulas differ from species to species. On the other hand, it is possible to produce light by mixing luciferin from one form and luciferase from another. The color of the light seems to be determined by the luciferase.

4. E. Newton Harvey, *Living Light* (Princeton: Princeton University Press, 1948), 58.

5. In the fluorescent bulb an electric discharge through mercury vapor produces a spectrum of emission lines, largely in the ultraviolet. The inside of the tube is coated with materials that fluoresce, producing visible light. One can choose different materials and thereby vary the spectral composition of the light. For example, the standard "cool-white" bulb confines most of its energy to the yellow-green region in the mid-visible, where human visual response is strongest. A "broad spectrum" bulb approximates the part of the sun's spectrum in the visible. In either case, fluorescence is cold light, producing little heat and using relatively little electricty.

6. To put it more precisely, quantum mechanics specifies probability distributions for the electrons. The two sigma electrons have the same probability distribution. It is concentrated strongly along the line between the two atoms. The electrons differ only in the direction of their spin. The two pi electrons have the same distribution, but it is a different distribution from that of the sigma electrons.

7. The terms *singlet* and *triplet* derive from consideration of the total spin of the two electrons. If the electrons have opposite spins, then the total spin is zero. (The two spins cancel.) Thus there is no such thing as a direction for this rotation; there is only one state, and it is called a singlet. When the two electrons have parallel spins, the total spin is double that of one electron. This total spin *can* vary in direction, and it turns out that quantum mechanics allows only three possible directions; so it is called a triplet. (The state of a single electron, in this nomenclature, is a *doublet.)*

8. Dyes were originally made from coal tar, a product of partially decayed formerly living matter. Now they are made synthetically.

9. There are some minor exceptions to these general statements. Certain bacteria process inorganic material (including carbon dioxide as a carbon source) for their growth. Certain groups of organisms near deep-sea hydrothermal vents live without reliance on the sun.

10. The words of Eugene I. Rabinowitch, one of the important contributors to twentieth-century research on photosynthesis. (*Scientific American* [August, 1948]: 25).

11. In the fall when chlorophyll in leaves fades, the remaining pigments are seen and the leaves turn yellow. Red pigments (not in-

volved in photosynthesis) are often newly manufactured in the autumn.

12. These techniques have since come to be used widely in biochemical research. A similar experiment was carried out as part of the Viking mission's search for life on Mars. A sample of Martian soil was exposed to radioactive CO_2, and after some time the soil was analyzed to see if microorganisms had metabolized the radioactive carbon.

Chapter 9

1. Data and estimates are from Earl Cook, "The Flow of Energy in an Industrial Society," in *Energy and Power* (San Francisco: Freeman, 1971), 83.

2. Let us think of "burning" as the oxidation of carbon leading to the formation of CO_2, which has two oxygen atoms for each carbon. Now a molecule like glucose, a sugar, has *one* oxygen atom for each carbon. We can think of it then as part of the way toward oxidation; it is intermediate between CO_2 and pure carbon (which we find in coal). Glucose is oxidized, but not as oxidized as CO_2. Ethanol, on the other hand, has one oxygen for every *two* carbons. Therefore it is less oxidized than glucose; it can produce more energy when it burns.

3. There are two types of coherence, one associated with moving *across* the light beam (spatial coherence) and one associated with moving along the light beam (temporal coherence). Spatial coherence is discussed above. Temporal coherence has to do with whether there is any phase relation between the wave at one point and the wave at another point downstream, along the direction of travel of the beam. The reason for a lack of temporal coherence is that each emission of light by an atom, in the sun for example, takes place only over a short period of time. The wave train that is produced by that emission does not stretch out over the entire distance from the sun to the observer; rather it stretches out only over a short distance, of the order of a centimeter. Coherence between parts of the beam separated by larger distances does not exist, because those parts of the beam have been emitted by different atoms, at unsynchronized times.

4. These are the heavy isotopes of hydrogen, used in these experiments because the temperature needed for their fusion is lower than that needed for fusion of ordinary hydrogen. Deuterium (hydrogen 2) is quite abundant in the oceans, essentially a limitless source of fuel. Tritium (hydrogen 3) is not found naturally and has to be manufactured from limited raw materials.

5. This is *amplitude modulation*, used by AM radio stations. *Frequency modulation* (FM) involves a slow variation of the radio frequency.

Chapter 10

1. *The Scientific Papers of James Clerk Maxwell* (New York: Dover Publications, 1965), 2:775.

2. Einstein does not refer to the Michelson-Morley experiment but takes Maxwell's theory as his starting point.

3. Here our discussion is restricted to light traveling through a

vacuum. Through a medium such as water or air the speed of light is smaller, as a result of interactions with the molecules of the medium, and we shall avoid dealing with these complications here. Propagation in vacuo brings us up against the nature of space and time.

4. The ether's function as the "substance" that waves is also unnecessary. Electric and magnetic fields exist in the vacuum, and their oscillations constitute what we perceive as light.

5. One way to visualize this rest energy is to picture the electron as a sort of ball of negative electric charge. (Its diameter would be about 10^{-15} meters.) If we imagine trying to assemble an electron from bits of negative charge widely spread about in space, we see that quite strong forces would have to be exerted to overcome the repulsive electrical forces that like charges always exert on each other. A particle constructed in this way would possess an internal tension, like a coiled spring, which we might characterize as a form of energy.

6. Modern accelerators can bring electrons up to a speed of more than 99.99999999 percent of the speed of light. To be precise, relativity implies that no object can be brought from a speed below c up to c or higher. It does not preclude the existence of particles that *always* travel at speeds higher than c. Physicists have speculated about such hypothetical particles (called tachyons), but no evidence has been found for their presence in nature.

7. That the photon's rest mass is precisely zero is, of course, to be determined by experiment. Because there are inevitably limits to the accuracy of any experiment, the present data consist of upper limits to the possible mass of the photon. The most recent data limit the photon rest mass to less than 10^{-48} grams; put another way, this mass is about 10^{-24} times the mass of a single hydrogen atom.

8. The following argument shows why the primitive process of a single electron emitting a photon cannot occur. Imagine the electron at rest. Its total energy is just its rest energy. Now suppose it emits a photon, which travels in some direction, say to the right. The electron recoils to the left. The total energy of the system now is the rest energy of the electron plus the energy of motion of the electron plus the energy of the photon; clearly more energy than existed at first. Thus a stationary electron cannot emit a photon, because energy conservation would be violated. Now consider a single moving electron. It is moving only relative to some arbitrarily chosen reference frame, and there is no reason why we cannot observe the electron in a frame of reference moving with it. In the new frame of reference the electron is stationary. It follows that to the observer in this frame it is impossible for the electron to emit a photon. Now if the laws of physics prevent the emission of a photon as observed in one frame of reference, it must be true that a photon cannot be emitted at all, because the laws of physics cannot depend on the frame of reference of the observer.

9. Quantum mechanics, we recall, imposes a limit on how accurately we can measure physical quantities. When energy is measured there is an uncertainty, and this uncertainty is inversely proportional to the time over which the measurement is made. Therefore, for a time interval short enough, we are granted a certain leeway in how well the

system's energy is known. The exchanged photon exists only within that margin of uncertainty.

10. Note also that the action-at-a-distance concept of interaction is inconsistent with the theory of relativity, because the concept requires that an influence created by one body be instantaneously felt by another body a distance away. Relativity insists that any influence can be propagated no faster than the speed of light. And so some type of intermediary (the field, quantized or not) traveling at finite speed must be part of any theory of interaction. Again we see that Maxwell's theory of light held the seeds of relativity.

11. Another important weak process is the conversion of a proton into a neutron, an antineutrino, and a positron. This process occurs under conditions of high density and is one of the steps in the nuclear burning of hydrogen in the center of a star.

Appendix D

1. Radioactivity was discovered in 1896; later, different components were distinguished by Rutherford and others.

Appendix E

1. A very recent survey suggests significant inhomogeneity in the distribution of galaxies through space.

2. Einstein's theory of relativity requires some slight modification (which we will omit) in this argument.

Appendix G

1. Band structure, of course, is more than a theory. It is supported by various types of experiments, among them studies of the interaction of light and ultraviolet with matter. The absorption of radiation by a crystal is found to occur over bands of wavelengths, corresponding to the excitation of electrons from one band to another (in contrast to absorption by gases, which occurs only at discrete wavelength values).

2. High temperature, in fact, lowers the conductivity of metals because vibrating nuclei get in the way of the flowing electrons.

Appendix I

1. Along the way it has been verified, with varying degrees of precision, that the speed of electromagnetic waves is the same over the entire spectrum, from long radio waves through the visible, and as far as gamma rays at the shortest wavelengths.

Index